INTERNAL AUDIT BRIEFINGS

INTERNAL AUDIT SAMPLING

Barbara Apostolou, PhD, CPA
and Francine Alleman, MS, CPA

The Institute of Internal Auditors

ISBN 0-89413-241-5
91033 5/91 h
93059 3/93 h

CONTENTS

ABOUT THE AUTHORS

Barbara Apostolou, PhD, CPA is Assistant Professor of accounting at Louisiana State University and an active member of The Institute of Internal Auditors. She is the author of numerous publications including other books for The IIA.

Francine Alleman, MS, CPA is a member of the professional audit staff of KPMG Peat Marwick. She has assisted on many projects for The Institute of Internal Auditors.

PART 1

INTRODUCTION TO AUDIT SAMPLING

1
INTRODUCTION TO AUDIT SAMPLING

Professional Internal Auditing Standards

Internal auditing is an independent and objective appraisal function within an organization. Its objective is to serve management and the board of directors in meeting organizational goals. The manner in which the internal auditing function performs its duties is governed by *Standards for the Professional Practice of Internal Auditing*, issued by The Institute of Internal Auditors. A detailed discussion of professional standards that impact audit sampling is presented in Chapter 5. However, a principal reason why internal auditors sample follows directly from the *Standards* in Section 420.01.2, *Examining and Evaluating Information*:

> Information should be *sufficient*, competent, relevant, and useful to provide a *sound* basis for audit findings and recommendations.
>
> *Sufficient* information is factual, adequate, and convincing so that a prudent, informed person would reach the same conclusions as the auditor.

Section 420.01.3 of the *Standards* also acknowledges that sampling techniques *may* be employed. The auditor should exercise professional judgment in deciding how much information is *sufficient* as well as whether it provides a *sound* basis for a conclusion. However, the *Standards* do not require that an auditor examine 100 percent of the information being evaluated. Nor do the *Standards* expect an absolute guarantee about the conclusions expressed.

Cost Versus Benefit Considerations

A second reason for selecting a sampling approach for evidence collection is due to time and cost considerations. The large size of most organizations precludes a 100 percent examination of underlying data. Even if 100 percent of the information could be tested, the cost of testing would likely exceed the expected benefits (the assurance that accompanies examining 100 percent of the total) to be derived.

Corroborating Information

The auditor is primarily concerned with corroborating representations made by the client or auditee. The emphasis is on whether the information being evaluated is "substantially or materially correct" rather than "absolutely accurate." In addition, auditors generally obtain evidence from multiple sources to use in appraising the reasonableness of representations so that they do not have to rely only on the sample results.

To summarize, internal auditors sample for three primary reasons:

1. The *Standards* permit sampling to obtain sufficient evidence to form a conclusion.
2. Sampling is often more efficient than a 100 percent examination.
3. Auditors express conclusions within materiality limits and are not expected or required to be 100 percent certain about their conclusions.

Written Policy Statement

Sampling is a generally accepted approach to accumulating audit evidence. An internal auditing department should prepare a written policy statement regarding sampling to guide auditors in the implementation of their sampling plans. This policy statement should address three issues at a minimum. Each is discussed separately below.

The policy statement should define *when* the circumstances are appropriate for statistical versus nonstatistical sampling. The policy can be general, such as "the use of statistical sampling should be considered, although it is not required." Alternatively, a more restrictive policy might be "statistical sampling is required in all situations where less than 100 percent of the items are being tested unless impossible or impractical to do so."

A second issue is to define *who* is authorized to design and implement sampling plans. This issue ensures that the internal auditing department is complying with *Standard 200: Professional Proficiency*. Some departments provide specialized training for certain individual internal auditors and designate them as "specialists" to supervise all sampling plans administered within the internal auditing function. As all members of the department gain experience and expertise in sampling applications, the specialist's role can be limited to unusual or complex sampling situations.

The third issue is to articulate *how* to implement a sampling plan in the written policy statement. This practice provides assurance that sampling plans are properly fashioned. Preprinted forms or pro forma working papers can be used to streamline a sampling plan and to document the process from planning through analysis of the results. The policy statement also should address how to deal with evaluation of the sample and potential problems, such as accounting adjustments, that may arise.

When Sampling Is Not Appropriate

One way to emphasize the propriety of sampling is to identify those audit situations in which sampling is clearly *not* appropriate. In general, three types of audit procedures fall into this category:

1. *Examination of 100 percent* of an account balance, class of transactions, or set of procedures. Clearly the auditor will not sample since the entire population is being audited. An example is confirmation of the entire balance of "loan payable" with a creditor.
2. *Inquiry and observation* are common procedures used to corroborate auditee representations. An example is an internal control structure questionnaire. The auditor will not select a portion, or sample, of items on the questionnaire to draw conclusions about the effectiveness of internal control procedures. Rather, it must be completed in full. Another example is a surprise count of teller cash. While sampling can be employed to select the *teller*, it is inappropriate to count a portion of the cash drawer and conclude about the total amount of cash in the drawer.
3. *Analytical procedures* include trend and ratio analysis and preparation of common size financial statements. These procedures involve the study of plausible relationships and actual outcomes to evaluate the appropriateness of values reported by the auditee. Since analytical procedures, by definition, examine the whole and its relationships to other measures, sampling is clearly inappropriate.

2
AUDIT RISK MODEL

Statement on Auditing Standards (SAS) No. 39, Audit Sampling[1] permits statistical or nonstatistical sampling to collect audit evidence. When an auditor samples, uncertainty enters the audit process. This uncertainty can be measured and controlled by using the audit risk model defined in *SAS No. 47, Audit Risk and Materiality in Conducting an Audit*[2]. The audit risk model is:

$$AR = IR \times CR \times DR$$

Where:

AR = **Audit Risk** = the risk that the auditor may fail to modify an opinion on financial statements that are materially misleading.

IR = **Inherent Risk** = the risk of material error in a transaction class or account balance assuming there are no internal controls.

CR = **Control Risk** = the risk that the internal control structure will fail to prevent or detect a material error.

DR = **Detection Risk** = the risk that the auditor's substantive tests will fail to detect a material error. Detection risk includes both tests of details and analytical procedures.

Audit risk includes uncertainties due to sampling and factors other than sampling. These aspects of audit risk are called *sampling risk* and *nonsampling risk*, respectively.

Sampling Risk

Sampling risk is the risk that the conclusions reached by the auditor based on an analysis of a sample will differ from those conclusions that would be reached by examining the entire population. In other words, the sample was not representative of the population. Typically the smaller the sample size, the greater will be the sampling risk. The auditor can use statistical sampling to control sampling risk. Sampling risk cannot be measured in a nonstatistical sampling plan.

Sampling risk affects both the efficiency and the effectiveness of audit procedures, as shown in Exhibit 2-1. The auditor is mainly concerned with the *effectiveness* components of sampling risk because they address the failure to detect a material misstatement or internal control weakness. However, the *efficiency* components indicate the auditor may expend more audit effort than necessary. This is true because the auditor will expand the scope of the audit when a sample indicates a control is ineffective or that a balance is misstated. The expanded scope should discover the "overassessment" or "incorrect rejection" errors indicated by the sample, while at the same time result in overauditing.

EXHIBIT 2-1
COMPONENTS OF SAMPLING RISK

	Efficiency	Effectiveness
Tests of Controls[3]	Risk of Assessing Control Risk Too High	Risk of Assessing Control Risk Too Low
Substantive Tests	Risk of Incorrect Rejection	Risk of Incorrect Acceptance
Statistical Term	Alpha Risk or Type I Error	Beta Risk or Type II Error

Nonsampling Risk

Nonsampling risk includes the risk the auditor will fail to detect a material internal control weakness or dollar misstatement due to factors other than sampling. Nonsampling risk arises even if the entire population is tested. Aspects of nonsampling risk include errors in auditor judgment such as:

- Application of an inappropriate audit procedure.
- Failure to recognize an error when one is present.

Nonsampling risk can be minimized in an audit by adhering to applicable professional standards. These standards include planning, supervision, and implementing quality control mechanisms.

Materiality

Materiality applies to substantive testing. The auditor should define how much monetary error can be tolerated in a transaction class or an account balance without resulting in financial statements that are materially misstated. The monetary error that can be tolerated is called *tolerable error*. Preliminary judgments about tolerable error are necessary in selecting an appropriate sample size. Materiality is defined for the overall financial statements and allocated over the accounts in a manner deemed appropriate by the auditor.

The estimate of materiality is inversely related to sample size. For example, if an auditor is unwilling to tolerate any monetary error, then 100 percent of the balance must be examined. If, however, tolerable error in an account is $50,000, the auditor can be satisfied by looking at less than 100 percent of the account. Assessment of materiality is a cost versus benefit decision.

3
STATISTICAL VERSUS NONSTATISTICAL SAMPLING

Audit sampling occurs whenever an auditor applies procedures to less than 100 percent of the total items being examined. The auditor may employ either statistical or nonstatistical sampling approaches. This section discusses the similarities and differences between each approach.

How to Distinguish Between Statistical and Nonstatistical Sampling

A sampling plan is *nonstatistical* when it fails to meet *at least one* of the criteria required of a *statistical* sampling plan. The auditor should learn the requirements of statistical plans because any deviation constitutes a nonstatistical approach. This book emphasizes statistical sampling. However, knowledge of the statistical requirements will permit the identification of nonstatistical applications as well.

Similarities Between Statistical and Nonstatistical Sampling

Three principal similarities exist between the two alternative approaches to sampling. First and foremost, both tactics require the exercise of auditor judgment. The auditor must exercise *judgment* during the planning, implementation, and evaluation stages of the sampling plan. In other words, the use of statistical methods does not eliminate the need to exercise judgment.

A second similarity is that the actual *audit procedures* performed on the items in the sample will be the same, whether a statistical or nonstatistical approach is used. The employment of a statistical plan does not mean the auditor can alter the procedures designed to collect evidence to draw an audit conclusion.

The third similarity between statistical and nonstatistical sampling is that *both are permitted* in practice. The IIA *Standards* (Section 420, *Examining and Evaluating Information*) note that evidence collected should support audit conclusions. The auditor must, therefore, exercise judgment in selecting a sampling approach to ensure compliance with the *Standards*. External auditing standards, namely *SAS No. 39* [4], permit either a statistical *or* a nonstatistical approach to sampling. Since both are allowed, the auditor should evaluate the relative costs and benefits associated with each before making a determination.

Differences Between Statistical and Nonstatistical Sampling Plans

Whenever an auditor draws a conclusion based upon examining only a portion of the total (sampling), the risk arises that the conclusion was incorrect. This is true simply because 100 percent of the items were not examined. Perhaps by chance the sample drawn was not representative of the total. This risk is *sampling risk.* Sampling risk can be measured and controlled in a statistical plan. Alternatively, even a perfectly designed *non*statistical plan cannot provide for the measurement of sampling risk. Risks associated with audit sampling are discussed in Chapter 2.

Additional differences between statistical and nonstatistical sampling plans are:

- *Technical knowledge* is required to implement a statistical plan. This additional knowledge can be obtained through training courses and programs coupled with experience.
- *Computer facilities* are required for most statistical applications because manual operations are cumbersome. However, any audit department with a personal computer owns the technology to handle most statistical applications.

Summary

Exhibits 3-1 and 3-2 present the similarities and differences, respectively, between statistical and nonstatistical sampling. The selection of which approach to use should be made after consideration of the following issues:

- Is an objective measure of sampling risk desired?
- What are the relative costs versus benefits involved?
- Is technical expertise available?

An additional consideration that favors using a statistical approach is that it provides objective conclusions rooted in mathematical theory. With the growing trend in litigation against auditors, both internal and external, a carefully documented statistical sampling plan is an excellent defense for conclusions expressed.

EXHIBIT 3-1
SIMILARITIES BETWEEN STATISTICAL AND NONSTATISTICAL SAMPLING

- Both Require Exercise of Auditor Judgment
- Audit Procedures Performed Will Not Differ
- Both Permitted in Auditing Practice

EXHIBIT 3-2
DIFFERENCES BETWEEN STATISTICAL AND NONSTATISTICAL SAMPLING

Statistical Plans:

- Control and Measure Sampling Risk
- Require Technical Training and Expertise
- Require Computer Facilities

4
TYPES OF STATISTICAL SAMPLING PLANS

Audit sampling can be performed on both monetary and nonmonetary populations. Using audit terminology, *tests of controls* are performed on nonmonetary populations. Alternatively, *substantive tests* are applied to monetary populations. Each topic is discussed in more detail in the remainder of this chapter.

Tests of Controls

An auditor performs tests of controls to assess whether they were functioning properly during the period being audited. Tests of controls call for the examination of documentation that offers evidence to determine if particular control procedures were performed. For example, an auditor may believe that a supervisory review of time cards (evidenced by signature) is a strong internal control procedure over the payroll cycle. A test of this control will confirm or disconfirm the auditor's belief. When performing tests of controls, the auditor measures compliance with the control procedure in terms of "error rates"— how often was the required control procedure *not* performed? Presence of control performance is called an **attribute**. Therefore, sampling in tests of controls is known as **attribute sampling**. Attribute sampling is covered in Part 4.

Substantive Testing

An auditor's objective in performing a substantive test is to evaluate the fair presentation of dollar amounts that appear in the financial statements. Two approaches to statistical sampling in substantive tests are probability-proportional-to-size sampling and variables sampling. **Probability-proportional-to-size (PPS)[5] sampling** is a variation of attribute sampling. PPS sampling gives greater weight to larger, more significant items and is useful for detecting overstatement when only a few errors are expected. This sampling approach is widely used in practice, especially for the audit of asset accounts. See Part 5 for a detailed discussion.

Variables sampling is used when the auditor wants to estimate a total dollar amount of transactions or balances. Three alternative approaches can be used, depending upon the specific circumstances (this topic is described more fully in Part 6):

1. **Mean per unit estimation** calculates the sample mean and multiplies it by the total number of items in the population to estimate total population value.

2. **Difference estimation** projects the average difference between book and audited values from the sample to the population.
3. **Ratio estimation** projects the ratio of audited to book values from the sample to the population.

Fraud Auditing

The auditor is occasionally confronted with a situation in which fraud is suspected, such as a suspicion that checks are being forged by an unscrupulous employee. A statistical sampling approach known as **discovery sampling** is used for this purpose. This sampling plan is designed so that if the irregularity does exist it will be discovered by the auditor. Hence the name, *discovery* sampling. A summary of statistical sampling plans is presented in Exhibit 4-1.

EXHIBIT 4-1
SUMMARY OF STATISTICAL SAMPLING PLANS

Audit Approach	Sampling Plan
• Tests of controls	• Attribute sampling
• Substantive tests	• PPS sampling • Variables sampling
• Fraud detection	• Discovery sampling

5
PROFESSIONAL STANDARDS GUIDING AUDIT SAMPLING

Sources of Standards

The Institute of Internal Auditors (The IIA) establishes *Standards for the Professional Practice of Internal Auditing*. The *Standards* establish the criteria by which the operations of an internal auditing department are evaluated and measured. The *Standards* are structured in this way:

- General standards (five).
- Specific standards (25) — must be complied with to follow general standards
- Guidelines (70) — state generally accepted guidance to meet the general and specific standards

The general and specific *Standards* are presented in Exhibit 5-1. Additional components of the *Standards*' framework include:

- *Statements on Internal Auditing Standards* (*SIAS*s) are authoritative interpretations of the *Standards*. *SIAS*s also may supersede or modify Guidelines.
- *Professional Standards Bulletins* (*PSB*s) are unofficial answers to questions raised about implementing the *Standards* in practice. They are chronicled by year. For example, *PSB 90-1* is the first unofficial question/answer issued in 1990.

An internal auditor must look to these sources when considering the use of sampling, whether it is statistical or nonstatistical.

How the Standards Address Sampling

The *Standards* do not specifically address sampling. *PSB 82-6* offers unofficial guidance and is reproduced in Exhibit 5-2. Essentially, *PSB 82-6* permits and encourages statistical sampling but *does not require* that it be used. Standard 420 is the official guidance that applies to audit sampling:

> **SS420.01.2** —Information should be *sufficient*, competent, relevant, and useful to provide a sound basis for audit findings and recommendations.
>
> *Sufficient* information is factual, *adequate*, and convincing so that a prudent, informed person would reach the same conclusions as the auditor.

Standard 420 provides the auditor with the latitude to exercise judgment in deciding whether to sample and what approach to use. In accordance with Standard 200, *Professional Proficiency*, the auditor has a duty to obtain the knowledge and skill that enters into this judgment.

The Certified Internal Auditor (CIA) examination tests the candidate's knowledge of statistical and nonstatistical sampling including:

- Background and perspective.
- Concepts and terms.
- Statistical and nonstatistical sampling.
- Types and plans.
- Sample selection.
- Sample and population evaluation.
- Practical problems.

The CIA designation is widely regarded as a sign of expertise in the profession of internal auditing. Its emphasis on sampling clearly demonstrates the importance that sampling bears to professional practice.

SIAS No. 5, Internal Auditors' Relationships with Independent Outside Auditors, issued by The IIA in 1987, highlights the importance of coordinating internal auditing work with the efforts of external auditors. Guideline 550.4 (of *SIAS No. 5*) calls for common understanding of audit techniques, methods, and terminology. More specifically:

> **G550.44** —It may be more efficient for internal and independent outside auditors to use similar techniques, methods, and terminology to effectively coordinate their work and to rely on the work of one another.

In the absence of more specific guidance on the use of sampling by internal auditors, *SIAS No. 5* implies that external auditing standards are an appropriate source of information for audit sampling practice. Refer to Chapter 2 for a summary of external auditing standards related to sampling.

EXHIBIT 5-1
SUMMARY OF THE IIA's GENERAL AND SPECIFIC *STANDARDS*

100 INDEPENDENCE
110 Organizational Status
120 Objectivity

200 PROFESSIONAL PROFICIENCY
The Internal Auditing Department
210 Staffing
220 Knowledge, Skills, and Disciplines
230 Supervision
The Internal Auditor
240 Compliance with Standards of Conduct
250 Knowledge, Skills, and Disciplines
260 Human Relations and Communications
270 Continuing Education
280 Due Professional Care

300 SCOPE OF WORK
310 Reliability and Integrity of Information
320 Compliance with Policies, Plans, Procedures, Laws and Regulations
330 Safeguarding of Assets
340 Economical and Efficient Use of Resources
350 Accomplishment of Established Objectives and Goals for Operations or Programs

400 PERFORMANCE OF AUDIT WORK
410 Planning the Audit
420 Examining and Evaluating Information
430 Communicating Results
440 Following Up

500 MANAGEMENT OF THE INTERNAL AUDITING DEPARTMENT
510 Purpose, Authority, and Responsibility
520 Planning
530 Policies and Procedures
540 Personnel Management and Development
550 External Auditors
560 Quality Assurance

EXHIBIT 5-2
PSB 82-6: PERFORMANCE — STATISTICAL SAMPLING

QUESTION: Is the internal auditor required to use statistical sampling when examining and evaluating the effectiveness of the organization's system of internal control?

ANSWER: Statistical sampling is *not* required when performing audit procedures. Nevertheless, sampling (statistical or judgmental[6]) is inherent in the auditing process. Sampling involves estimating the rate of occurrence of a particular characteristic or estimating population totals by examining less than all of the items in a population. Internal auditors are constantly faced with the assignment of drawing conclusions about entire populations based on the examination of only a portion of the items in the population. The decision to use statistical or judgmental sampling to select items for examination is based on the professional judgment of the auditor. Statistical sampling is useful for quantifying the judgment of the auditor.

In some circumstances statistical sampling is more appropriate than judgmental sampling. Before deciding whether to use statistical or judgmental sampling, the auditor must determine the audit objectives; identify the population characteristics of interest; and state the degree of risk that is acceptable. After making those determinations, it may be advisable to use statistical sampling if the auditor has a well-defined population and can easily access the necessary documentation.

REFERENCES

[1] New York: American Institute of Certified Public Accountants, 1981.

[2] *SAS No. 39* presents an audit risk model nearly identical to the one defined in *SAS No. 47* with one exception— the terminology. The *SAS No. 47* model is presented here because it is more commonly employed in practice.

[3] The components of sampling risk in tests of controls were formerly referred to as the *risk of underreliance* (alpha) and the *risk of overreliance* (beta). However, since the audit risk model calls for the assessment of control risk versus a rely-no rely decision, the risks are now expressed in terms of control risk assessment.

[4] New York: American Institute of Certified Public Accountants, 1981.

[5] Also known by the following terms:

- Dollar unit sampling (DUS).
- Combined attribute variables (CAV) sampling.
- Cumulative monetary amount (CMA) sampling.
- Monetary unit sampling (MUS).

[6] Referred to as *nonstatistical* sampling in this book.

ENDNOTES

PART 2

BASIC STATISTICAL CONCEPTS

6
POPULATION AND SAMPLE

Population

The total collection of items about which the auditor will express a conclusion is the *population*. A typical audit procedure is a test to determine whether checks were signed by authorized signers. In this case, the population is "all checks written for the period under review." The true answer to this audit objective is unknown and only can be determined if 100 percent of the checks are examined by the auditor.

Sampling allows the auditor to conclude, or draw an inference, about the population by examining only a portion of it. The auditor must exercise caution in defining the population because the results can be projected only to the population from which the sample items were drawn. For example, if the auditor draws a sample from checks written on the Blue Bank account, the conclusion reached could not be made about checks written on the Red Bank account since the Red checks were not part of the population sampled.

Sampling Unit

The individual items that make up the population are *sampling units*. These are the items to which the audit procedures will be applied. For example, when auditing accounts receivable, the sampling units could be customer accounts *or* individual sales invoices.

Frame

The sample *frame* is a listing of the sampling units that comprise the population. In the audit of accounts receivable, for example, the frame is a listing of the trial balance from the subsidiary accounts receivable ledger. Development of an adequate frame is a requirement for the sample selection process.

Sample

A *sample* is the collection of sampling units drawn from the frame that will be subjected to audit procedures. The inferences drawn about the population will be based on the evidence collected from the sample.

7
MEASURES OF LOCATION

Population and sample data are often described in terms of a representative value, called a measure of location, or central tendency. There are three measures of location — mean, median, and mode. The **mean**, or average, is computed by taking the total value of the population (sample) and dividing by the number of items in the population (sample). The mean is also the expected value of an item in the population (sample). Population and sample means can be computed with the following formulas:

Population	Sample
$\mu = \sum_{i=1}^{N} x_i \div N$	$\bar{x} = \sum_{i=1}^{n} x_i \div n$

Where: $\sum_{i=1}^{N(n)} x_i$ = sum of the items in the population (N) or sample (n)

N = number of items in the population

n = number of items in the sample

μ = population mean

$\bar{x}$ = sample mean

The **median** is the middle value in a population (sample). It is computed by ordering the values from lowest to highest and selecting the "middle" value. The **mode** is the most frequently occurring value in a population (sample).

Example Computations

Suppose that accounts payable consists of the following balances:

1	$ 325	
2	390	← **mode**
3	390	← **mode**
4	755	← **median** and **mean**
5	793	
6	1,069	
7	1,563	
	$ 5,285	

- Computation of population *mean*, μ:

$$\mu = \sum_{i=1}^{N} x_i \div N$$

$$\mu = \$5{,}285 \div 7$$

$$\mu = \$\ 755$$

8
MEASURES OF VARIABILITY

A population (sample) is often described in terms of how its individual items vary in value. Several different measures can be used to describe this variation within the population (sample). The *range* is the difference between the lowest and highest values in the population (sample). The *variance* measures the variation about the mean value in the population (sample).

Population

$$\sigma^2 = \frac{\Sigma (x_i - \mu)^2}{N}$$

Sample

$$s^2 = \frac{\Sigma (x_i - \bar{x})^2}{n-1}$$

Where: σ^2 = population variance
s^2 = sample variance
x_i = value of ith item
μ = population mean
$\bar{x}$ = sample mean
N = number of items in the population
n = number of items in the sample
Σ = summation operation

The *standard deviation* is the square root of the variance. Population standard deviation is denoted by and sample standard deviation by s.

Example Computations

Accounts payable consists of the following balances (μ = $755):

	x_i	$x_i - \mu$	$(x_i - \mu)^2$
1	$ 325	(430)	184,900
2	390	(365)	133,225
3	390	(365)	133,225
4	755	0	0
5	793	38	1,444
6	1,069	314	98,596
7	1,563	808	652,864
	$5,285		$ 1,204,254

Variance:

$$\sigma^2 = \frac{\Sigma(x_i - \mu)^2}{N} = \frac{1{,}204{,}254}{7} = \mathbf{172{,}036}$$

Standard Deviation:

$$\sigma = \sqrt{\sigma^2} = \sqrt{172{,}036} = \mathbf{415}$$

Range:

$$(1{,}563 - 325) = \mathbf{1{,}238}$$

Confidence

The confidence level is used in statistical sampling to describe the degree of belief the auditor has in the obtained results. It is expressed in terms of a percentage. To be *100 percent* confident is to be *certain* of the results. This outcome is unrealistic in sampling situations. As a result, most auditors seek a confidence level of *at least 90 percent* in describing sample results. For example, *95 percent* confidence is often used in auditing. This means that if 100 samples of **n** items were taken and 100 confidence intervals were constructed, 95 of them would contain the true value of the population. A *confidence interval* describes the range of likely values within which the true population value should lie. It is constructed by taking the **sample mean ± some margin of error**, which is discussed in Part 6. The recorded population amount is compared to the values in the confidence interval. If the recorded value lies within the interval, then the auditor can be *95 percent*[7] confident the recorded balance is not misstated by a material amount.

Precision

Precision intervals are the same as confidence intervals. *Precision* measures how close the sample estimate is to the true but unknown population value. For example, an auditor computes *precision* of $7,500 in the audit of an account with a balance of $180,000. The *precision interval* is $180,000 ± $7,500, or $172,500←→$187,500. Precision can be *planned* (what the auditor desires) and *achieved* (computed based on sample results).

Reliability

Reliability is a term used in audit sampling to refer to "confidence level." If sample results are reported at a reliability of .95 or 95 percent, the auditor is 95 percent confident in the conclusions.

REFERENCES

[7] Or the confidence level specified by the auditor.

PART 3

SAMPLE SELECTION METHODS

9
STATISTICAL SAMPLE SELECTION

Once the auditor decides to implement sampling, a sample selection method must be chosen. If statistical sampling is being used, the method may be dictated by the particular technique employed. The determination also will depend upon the nature of the data being sampled.

Statistical Sample Selection Methods

Valid statistical inferences can be made only when each item in the population has a known chance, or probability, of being selected. Two widely used approaches that meet this criterion are:

- Simple random sampling.
- Systematic sampling.

Simple Random Sampling

Simple random sampling requires that all items in the population have an equal chance of being selected. In auditing applications, this method uses sampling without replacement; that is, once an item has been selected for testing it is removed from the frame and is not subject to re-selection. An auditor can implement simple random sampling in either of two ways:

- Computer Programs.
- Random Number Tables.

Computer Programs

Computer programs are commonly available to select random samples. The only input required from the auditor is a consecutive ordering of the population. For example, if the auditor wishes to audit cash disbursements, a random sample of checks is selected. Assume that checks issued for the period were numbered from **3942** to **5706**. Given the start and stop points for the check sequence, the program will randomly select the specific check numbers to be audited.

If the population of interest is not conveniently numbered as in the check example, the auditor may do so. Some documents are identified using an alphanumeric scheme. For example, customer accounts may be identified as such:

H100 - H750 Sales Region "H"
N100 - N400 Sales Region "N"

The auditor can easily convert these accounts into a consecutively numbered population by assigning the number "1" to replace "H" and the number "2" to replace "N." The input becomes:

1,100 - 1,750 Sales Region "H"
2,100 - 2,400 Sales Region "N"

The program can now randomly select the required sample from this population. The auditor must convert it back to identify the actual customer account —1,309 translates to "customer account H309."

The auditor may encounter unnumbered populations from which a random sample is desired. If this occurs, the auditor should develop a numbering technique that is convenient. If the items appear on a computer printout, they could be numbered by page and line number to create a consecutive sequence. If a population is in such disarray that no reasonable conversion scheme can be developed, the auditor should consider abandoning random selection in favor of a more suitable approach.

Random Number Tables

For those auditors without access to computer programs, a random sample can be selected by use of random number tables. This task is easily structured into three basic steps:

1. Define the correspondence of the population to the values in the random number tables.
2. Determine the direction or path to take through the tables in selecting items.
3. Pick a random start in the tables. This step can be accomplished by dropping a pencil point on the page while looking away — the so-called *random stab* approach.

The auditor's working papers should clearly document these three steps. Refer to Exhibit 9-1 for an illustrative working paper.

EXHIBIT 9-1
RANDOM NUMBER SELECTION USING
RANDOM NUMBER TABLE

7035	3543	1022	9449	5954 Random	7962
3674	8138	4948	787	3839 start	[4855] (1)
→ 1549	1576	4699 (2)	7592 (3)	6505 (4)	3858 (5)
3663	8046	8098	5218 (6)	1408	2062
1547	4291 (7)	933	8257	8909	5957 (8)
646	8421	5512 (9)	9862	9315	5324 (10)
9668	6892 (11)	4733 (12)	4543 (13)	9485	8419
6679 (14)	1210	6970 (15)	8508	6817 (16)	4999 (17)
7273 (18)	8424	9594	4094 (19)	933	6275 (20)
2623	1967	1440	7378 (21)	5874 (22)	3328
7726 (23)	1923	7027 (24)	499	8109	8079
2663	6254 (25)	7859	2780	8184	3344
2170	7079	8593	7200	6310	1069
1348	8549	8419	1269	4712	278
7816	5386	433	3900	2050	8819
3121	3652	4343	2298	7721	5462

Population: Invoice # 3768 - 7592
Sample Size: 25 invoices

Route: (1) Circle random start.
(2) Move horizontally from random start, left to right.
(3) Go down to left side of next row and move right.

This example shows the route taken through the table and which items were selected.

Systematic Sampling

Systematic sample selection can be applied to the physical units in a population (such as checks, accounts, inventory items) or to the monetary units (dollars). When applied to the monetary units, each dollar is the sampling unit. The physical unit containing the dollar is then audited. Systematic sampling with monetary units is a special case of probability-proportional-to-size (PPS) sampling. A thorough presentation of this sampling technique is made in Part 5 of this book. This chapter deals with systematic sampling of physical units. The auditor first calculates the *sampling interval:*

$$\text{Sampling Interval} = \frac{\text{Population Size}^{8}}{\text{Sample Size}}$$

The value of the sampling interval is always rounded down to ensure that the desired sample size is achieved. For example, the sampling interval for a population of 1,512 items and required sample size of 60 is *25* (1,512 ÷ 60 = 25.2).

The second step in systematic sampling is to compute a random start within the first sampling interval; that is, items 1-25. Including the random start, every twenty-fifth item is selected for inclusion in the sample. Some auditors guard against potential bias in the selection process by using more than one random start. If this goal is desired, then the sampling interval is multiplied by the number of random starts. For example, if three random starts are used in the illustration above, the sampling interval is 75 (3 times 25). Three random starts are generated within the first interval of 1-75, with every seventy-fifth item selected from each start.

Systematic sampling is typically easier to employ than random sampling. However, it may produce a nonrepresentative sample if the population contains a systematic bias. The auditor can guard against bias by ensuring that the population is in random order, which fortunately applies to most audit populations. Using multiple random starts is also an excellent tactic. In fact, many auditors employ multiple random starts as a matter of policy. With appropriate attention to the potential for bias, the auditor can use systematic sampling in statistical applications.

10
NONSTATISTICAL SAMPLE SELECTION

Statistical sample selection methods also can be used for nonstatistical sampling plans. Some selection methods, to be discussed here, can be used only with nonstatistical sampling plans. They include haphazard, block, and judgment selection.

Haphazard Selection

In this method, the auditor selects the sample items without intentional bias to include or exclude certain items in the population. It represents the auditor's best estimate of a representative sample — and may in fact be representative. Defined probability concepts are not employed. As a result, such a sample may *not* be used for statistical inferences. Haphazard selection is permitted for nonstatistical samples when the auditor believes it produces a fairly representative sample.

Block Selection

Block selection is performed by applying audit procedures to items, such as transactions, all of which occurred in the same "block" of time or sequence of transactions. For example, in testing sales, the auditor may examine all invoices issued in the first week of November. Alternatively, invoices **3001 - 3100** may be examined in their entirety. Block selection should be used with caution because valid inferences cannot be made beyond the period or block examined. If block sampling *is* used, many blocks should be selected to help minimize sampling risk.

Judgment Selection

Judgment sample selection is based on the auditor's sound and seasoned judgment. Three basic issues determine which items are selected:

1. *Value of items.* A sufficient number of high-dollar items should be included to provide adequate audit coverage.
2. *Relative risk.* Items prone to error due to their nature or age should be given special attention. Examples include old unpaid receivables or payables, transactions arising in a weak internal control structure, or complex transactions.

3. *Representativeness.* Besides value and risk considerations, the auditor should be satisfied that the sample provides breadth of coverage over all types of transactions or balances in the population.

Stratified Sampling

Stratification is the process of separating the population of interest into sub-populations called strata. Samples are then selected from each stratum. The individual strata should consist of relatively homogeneous sampling units, as measured by the standard deviation (σ). The purpose of stratified sampling is to achieve a reduced sample size by reducing variation within each sub-population. Stratified sampling can be employed in both statistical and nonstatistical sampling plans. For nonstatistical plans, variation is measured in qualitative terms (small, moderate, large) rather than with the standard deviation (σ). For a statistical plan, the σ can be estimated by using the results of a pilot sample — usually 30-50 items per stratum. The pilot sample can be designed so that the items can be included in the main sample as well. An alternative approach is to use prior period audit results to avoid the costs involved with selecting a pilot sample.

11
CONSIDERATIONS IN SELECTING A SAMPLE

Regardless of the sample selection method employed or whether a statistical or nonstatistical approach is used, the auditor must deal with unusual issues from time to time. Two of the more familiar situations and suggested remedies are identified below:

1. *Missing Items.* What should the auditor do when an item to be sampled cannot be found? In tests of controls, the missing document (check, invoice, etc.) cannot be audited and must be treated as a control deviation or finding. One outcome is that control risk is assessed at a higher level than planned and the substantive audit plan must be revised. For internal auditors, however, the most likely outcome is that the presence of the missing item(s) will be highlighted in the audit report.

   When supporting documentation cannot be located to support a dollar balance in a substantive audit test, the auditor must assign the item an audited value of $0 (zero). This result could lead to rejection of the book value and the expenditure of additional audit effort. Clearly all effort should be made to locate missing documents.
2. *Voids.* Documents are voided for various reasons in the ordinary course of business. When a voided document, such as a check or sales invoice, is encountered, it should be replaced[9] with a new item. Audit procedures are then applied to the replacement item. If a substantial number of voids are discovered, the auditor may wish to investigate and report on the cause as a separate matter.

REFERENCES

[8] The population is expressed in either units or dollars, depending upon the objective.

[9] Most auditors select a few items over and above the required sample in the event replacement items are needed.

PART 4

ATTRIBUTE SAMPLING

12
OBJECTIVES OF TESTS OF CONTROLS

Internal auditors frequently sample for the purpose of drawing conclusions about the *effectiveness* of control procedures. These procedures are known as tests of controls. A control procedure is defined as a qualitative feature or *attribute* of the items contained in the population. Generally, the auditor views control effectiveness based on the frequency with which the control policy or procedure *did not occur*. In other words, how often did a procedure *deviate* from those prescribed by management to meet established control objectives? The evaluation typically involves a *yes* or *no* decision—either the control procedure was or was not performed properly.

Attribute sampling plans are appropriate whenever the auditor is testing for the effectiveness of a prescribed control procedure (tests of controls). In general, there are seven principal steps to be conducted in an attribute sampling plan as summarized in Exhibit 12-1. All seven steps should be clearly documented in the audit working papers.

EXHIBIT 12-1
STEPS IN CONDUCTING AN ATTRIBUTE SAMPLING PLAN

1. Determine the objectives of the audit test.
2. Define the attribute *and* deviation condition(s).
3. Define the population.
4. Determine the sample selection method.
5. Determine the sample size.
6. Perform audit procedures on the sample items.
7. Evaluate the sample and express conclusion(s).

Control procedures are established by management to provide assurance that organizational objectives are achieved. One objective, for example, is to ensure that the accounting records are free of material misstatement. A control procedure may be preventive or detective in nature, depending upon managerial assessment of the corresponding costs and benefits. The auditor's *objective* in an attribute sampling plan is to evaluate the effectiveness of, or compliance with, a prescribed control procedure.

13
FACTORS THAT IMPACT SAMPLE SIZE

The four factors that impact sample size in attribute sampling plans are summarized in Exhibit 13-1. These factors must be considered in combination with one another to achieve the required sample size.

Population Size

In general, the larger the population, the greater the sample size. However, most audit applications assume a "large" population of over 5,000 sampling units. In this case, population size has a minimal impact on sample size. In fact, little difference is noted in sample sizes except where a population contains less than 500 items.

Risk of Assessing Control Risk Too Low

If an auditor assesses control risk too low, then a danger exists that the audit will be ineffective[10]. See Chapter 2 for a discussion of this aspect of sampling risk. An auditor typically establishes this risk (beta risk) in the neighborhood of five percent (.05). If beta risk is decreased to a smaller amount, say three percent (.03), then the auditor would be required to examine more items to obtain the increased confidence corresponding to the lower risk assessment. Alternatively, if the risk of assessing control risk too low is set at a higher value, say 10 percent (.10), then fewer items would need to be tested because less confidence is desired in the results. Since the required sample size moves in the direction opposite to changes in the risk of assessing control risk too low, this sample size factor has an *inverse* impact on sample size.

Tolerable Deviation Rate

When performing tests of controls, the auditor must specify the maximum number of deviations from the prescribed control procedure that can be tolerated without affecting acceptance of the procedure as effective. This measure is known as the *tolerable deviation rate* or occurrence rate. For example, an auditor may be willing to tolerate a four percent (.04) deviation rate. If this rate were increased to eight percent (.08), what is the impact on sample size? Clearly, the sample size would be smaller. Alternatively, if the tolerable deviation rate were lowered to two percent (.02), then many more items would need to be examined. As a result, tolerable deviation rate has an *inverse* impact on sample size.

Expected Population Deviation Rate

The auditor must estimate what the expected population deviation rate is prior to performing tests of controls. This estimate is the auditor's best guess of how often a control procedure was not performed when it was prescribed and is based on experience, inquiry, and observation. If the expected population deviation rate *exceeds* the tolerable deviation rate, then tests of controls are not indicated! In this event, the auditor should focus on substantive audit procedures that evaluate the effects of the absence of the prescribed control procedure. In addition, this case can be a fertile source of findings for inclusion in the internal audit report.

However, in situations where the expected population deviation rate does *not* exceed the tolerable deviation rate, tests of controls are indicated. The impact of this factor on sample size is direct. That is, as the expected population deviation rate increases, the required sample size becomes larger. The reason for this result is that as the expected population deviation rate approaches the predefined tolerable deviation rate, the auditor requires more precise information to draw a conclusion. This increased precision calls for the examination of more sample items. Hence, the impact of expected population deviation rate on sample size is *direct*.

EXHIBIT 13-1
FACTORS THAT IMPACT SAMPLE SIZE IN ATTRIBUTE SAMPLING

Factor	Impact on Sample Size
• Population Size	• Direct
• Risk of Assessing Control Risk Too Low	• Inverse
• Tolerable Deviation Rate	• Inverse
• Expected Population Deviation Rate	• Direct

Direct = as sample size factor increases (decreases), sample size becomes larger (smaller).

Inverse = as sample size factor increases (decreases), sample size gets smaller (larger).

14
ATTRIBUTE AND DEVIATION CONDITIONS

An attribute is the control procedure and a deviation is the absence of the prescribed control procedure. Attribute sampling deals with a *yes* or *no* evaluation—either the control procedure is present or it is not. Both the attribute and the deviation condition(s) must be carefully defined and documented for the benefit of the auditors performing the audit procedures. Consider the following example:

Suppose an organization has a policy of requiring two authorized signatures on all checks issued. This policy is established to prevent unauthorized access to the organization's asset, cash. The auditor wants to test this control because it will provide assurance that disbursements have been authorized in accordance with management's objectives. The required steps are:

1. Define the attribute:
   - Two signatures as appearing on the depository bank's signature card.
2. Define the deviation condition(s):
   - Only one authorized signature.
   - No signatures.
   - Two signatures with one unauthorized.
   - Two unauthorized signatures.
   - Presence of any unauthorized signature even if in combination with the required authorized signatures.
   - The check is missing and cannot be located.

15
DEFINING THE POPULATION

An auditor samples to draw conclusions about *a particular* population. The sample results can be projected *only* to the population from which the sample items were drawn. If an auditor selects a sample of checks from Blue Bank to examine for the presence of dual signatures, the results apply *only* to checks issued on the account at Blue Bank and not to any other bank's checks. For example:

Bank	Number of Checks Issued
• Blue Bank	4,056
• Green Bank	9,468
• Orange Bank	10,870
Total Population	24,394

The population is defined as consisting of 24,394 sampling units—all checks drawn on all accounts during the period under audit. If the sample is selected so that checks from all three banks are tested, then the sample results can properly be projected to the population consisting of three banks and 24,394 checks.

16
METHODS OF SAMPLE SELECTION

Sample selection methods are described more fully in Part 3. A *statistical* attribute sampling plan will most likely employ simple random selection using a computer program or random number tables. These sample selection methods can be *fixed* or *sequential*. Under a fixed approach, the auditor selects a single sample of a mathematically computed size. A sequential plan calls for selecting the sample in several steps, with each step dependent upon the audit results of the previous step. The decision to use fixed or sequential sampling depends upon the auditor's assessment of which approach will provide the most efficient sample size. Sequential attribute sampling is presented in Chapter 19.

Prior to presenting the details of fixed versus sequential sampling plans, presentation of some statistical theory is appropriate. In performing tests of controls, the auditor must decide exactly what information is desired. The choice is between the following two situations, also summarized in Exhibit 16-1. The auditor should select carefully because the approach will determine which formulas or tables to use for sample size computations as well as for sample results evaluation.

Situation A

The auditor desires an estimate of the actual population deviation rate of a particular control procedure. The statistical result is a *two-sided precision interval*. The conclusion can be expressed in this way:

> We are 95 percent confident that the actual population deviation rate lies between three percent and five percent.

This conclusion is compared to what the auditor considers to be a tolerable deviation rate. Three possibilities arise:

1. If the tolerable deviation rate is higher than the upper precision limit (5 percent in the example), then the auditor may conclude that the control is functioning as prescribed.
2. If the tolerable deviation rate is *less* than the *lower* precision limit (three percent in the example), then the auditor may conclude that the control is *not* functioning as prescribed.[11]

3. If the tolerable deviation rate falls within the precision interval, say four percent for this example, then no conclusion can be expressed about the effectiveness of the control procedure. If this situation occurs, the auditor will be required to expand the scope of the audit in some fashion, such as increasing the sample size.

Situation B

The auditor wants to know whether the actual population deviation rate exceeds the tolerable deviation rate. The statistical result is a *one-sided upper precision limit*. The conclusion can be expressed this way:

> We are 95 percent confident that the actual population deviation rate does not exceed the tolerable deviation rate of five percent.

The *one-sided upper precision limit* approach is used by internal auditors in quality control audits. It permits the auditor to conclude at a stated level of confidence whether deviations exceed a predetermined tolerable deviation rate. The American Institute of Certified Public Accountants (AICPA) describes the *one-sided* approach in its Audit and Accounting Guide entitled *Audit Sampling* (1983). External auditors typically do not perform tests of controls unless they believe the control procedure is strong and functioning as prescribed. Thus, they are trying to determine whether the actual deviation rate is less than the tolerable deviation rate or not. No emphasis is placed on learning the exact "value" of the actual population deviation rate—only that it does not exceed the tolerable rate. Since the *one-sided* approach tells the auditor whether or not to *accept* the population as having less than the tolerable deviation rate, it is also known as *acceptance sampling*.

EXHIBIT 16-1
ALTERNATIVE CONCLUSIONS IN ATTRIBUTE SAMPLING

Situation	Statistical Term
• Estimate actual population deviation rate.	• Two-sided precision interval
• Does actual population deviation rate exceed tolerable deviation rate?	• One-sided upper precision limit

17
DETERMINING THE SAMPLE SIZE

In a fixed sampling plan, a single sample is selected on which audit procedures will be performed. The sample size can be computed by using a formula or by reference to convenient tables. The appropriate sample size formula for the two-sided precision interval approach is as follows:

$$\text{Sample Size} = n = \frac{z^2 \cdot pq}{(pre)^2}$$

Where: z^2 = z value that corresponds to the desired confidence level[12]

p = tolerable deviation rate[13]

q = $1 - p$

pre = tolerable deviation rate minus expected population deviation (precision)

Example

The auditor expects the population to have a deviation rate of four percent, but is willing to tolerate a deviation rate of five percent. The desired precision is ± one percent. The desired risk of assessing control risk too low is five percent, which translates to a z value of 1.96 (per Appendix E, Table 1). The sample size is **1,825**, computed as follows:

$$n = \frac{1.96^2 \times .05 \times .95}{(.01)^2}$$

$$n = \frac{.182476}{.0001}$$

$$n = \mathbf{1{,}825}$$

An alternative to the use of the sample size formula above is reference to the convenient tables presented in Appendix A. Suppose the following:

Population	10,000
Confidence Level	95%
Risk of Assessing Control Risk Too Low	5%
Expected Population Deviation Rate	4%
Desired Precision	± 1.25%
Tolerable Deviation Rate	6%

The appropriate sample size can be found in Appendix A, Table 2. Note that there are three tables for computing sample sizes—a, b, and c—that correspond to a 95 percent confidence level as established in the above example. Locate the table that refers to a four percent expected population deviation rate—Table C. Now find the value where population (10,000) and precision (± 1.25%) intersect on the table. The value, or sample size, is **283.** Note that computer programs are available (or can be written) to generate sample sizes as well.

When using the one-sided upper precision limit approach, the easiest alternative is to use published tables. They are easily employed by identifying the four sample size factors discussed previously. Consider this example:

Population	"large"
Tolerable Deviation Rate	4%
Expected Population Deviation Rate	1.5%
Risk of Assessing Control Risk Too Low	5%

Refer to Appendix B, Table 1, which provides sample sizes for situations where the risk of assessing control risk too low is five percent, as in this example. Appendix B, Table 2 offers sample sizes where this same risk is set at 10 percent. In Appendix B, Table 1 locate the intersection of the expected population deviation rate of 1.5 percent (left hand column) and the tolerable deviation rate of 4 percent (across top). The required sample size is **192** items. The **(3)** indicates the number of deviations in the sample that will permit the auditor to accept the control as being effective.

If the exact values are not in the table, the auditor can select the more conservative sample size or interpolate[14]. Alternatively, computer software is available to compute exact sample sizes[15]. Internal auditors who frequently use attribute sampling should invest in a computer program.

18
EVALUATING SAMPLE RESULTS

The auditor must perform audit procedures on the sample items after the sample is selected. This step calls for the auditor to complete the procedures identified in the audit program. In tests of controls, the auditor notes compliance with a prescribed control procedure and documents all instances of noncompliance (deviations as defined in Chapter 14). The audit work should be summarized and documented in the working papers.

Evaluate the Sample Results and Express Conclusion(s)

Two-sided Precision Intervals

Evaluation of sample results in a test designed to produce a two-sided precision interval can be accomplished by referring to Tables 3 through 7 located in Appendix A. Recall the situation presented in the previous chapter:

Population	10,000
Confidence Level	95%
Risk of Assessing Control Risk Too Low	5%
Expected Population Deviation Rate	4%
Desired Precision	± 1.25%
Tolerable Deviation Rate	5%

Suppose the sample results are:

Sample size	201
Sample deviation rate	2%

Refer to Appendix A, Table 4, which is for results yielding a sample deviation rate of 2 percent. Locate the appropriate confidence level (95 percent) and sample size (≈200). The two-sided precision interval is 0.6 ←→ 5.0. This interval can be expressed in the form of an audit conclusion:

> We are 95 percent confident that the actual population deviation rate lies between 0.6 percent and five percent. Since this is *less* than the tolerable deviation rate of five percent, the control procedure appears to be functioning as prescribed.

One-Sided Upper Precision Limit

Tables 3 and 4 in Appendix B also permit the auditor to evaluate the sample results and project them to the population. The table with the appropriate risk of assessing control risk too low should be used—Table 3 (five percent) or Table 4 (10 percent). Suppose the following:

Risk of Assessing Control Risk Too Low	5% (Table 3)
Sample Size[16]	163
Actual Number of Deviations	2

Refer to Table 3 and note that the intersection of 163 (Sample Size) and 2 (Actual Number of Deviations) is between 3.2 and 4.2. How should this result be interpreted? The auditor can choose the more conservative value (4.2) or can interpolate on a straight-line basis for a more precise estimate:

$$4.2 - [(4.2 - 3.2) \div (200 - 150) \times (163 - 150)]$$

$$4.2 - 0.26 \approx 3.9$$

This value can be interpreted as such:

> We are 95 percent confident that the true (but unknown) population deviation rate does not exceed 3.9 percent, which is less than the tolerable deviation rate of five percent.

So, the auditor concludes the control is effective and functioning as prescribed.

19
SEQUENTIAL ATTRIBUTE SAMPLING

Refer to Appendix B, Table 1. Notice that as the expected population deviation rate increases, the sample size also increases. For example, consider a tolerable deviation rate of five percent. When the expected population deviation rate is zero, the sample size is **59**. Notice that the required sample size more than doubles when the expected population deviation rate exceeds 1.25 percent. In fact, the tables do not give values for sample size once the expected population deviation rate exceeds 2.25 percent because they are too large to be efficient for audit purposes. Fixed sampling plans have little tolerance for expected deviations in the population—the sample sizes rapidly become too large to justify statistical sampling from a cost versus benefit perspective.

An alternative approach to fixed attribute sampling is sequential attribute sampling. The sequential approach is sometimes referred to as **stop-or-go sampling**. It is designed to tolerate a small number of errors efficiently while still providing the auditor with a statistical conclusion. A sequential sample includes a number of blocks of sample items and each block is audited in order. Sequential plans consist of an initial sample size, incremental sample blocks, and the total number of blocks possible. Not all blocks need to be sampled; the inclusion of each block is conditional on the outcome of the audit procedures performed on the previous block. Once a given block has been completed, the auditor stops and applies the following rules of thumb:

(a) How many deviations were noted?
(b) How many blocks have been audited?

If (a) is less than (b), then the desired conclusion has been obtained. However, if (a) is greater than or equal to (b), the auditor must decide whether to continue sampling or abandon the sampling plan. Criteria for this choice will be presented shortly.

Sample Size

Refer to Appendix C for tables designed to compute sample sizes in sequential sampling plans. Frequent users should obtain computer software for this purpose. The auditor must specify three variables to compute sample size (values are presented for illustrative purposes):

Beta Risk[17]	5%
Alpha Risk[18]	5%
Tolerable Deviation Rate	7%

Refer to Table 1(a) in Appendix C, which provides sample sizes for alpha risk of five percent. Locate the intersection of beta risk (left hand side) and tolerable deviation rate (across top). The initial required sample size is **44**. The incremental sample sizes (bottom line of table) are **40**.

Evaluating Sample Results

Refer to Table 2 in Appendix C. Sequential sampling requires the auditor to make a decision at the end of each block: **stop**, **continue**, or **abandon** the sample. The circumstances for each decision are:

- *Stop:* When the cumulative number of deviations is less than the number of blocks sampled, the auditor can conclude that the population deviation rate is less than the tolerable deviation rate at the specified alpha and beta risks.
- *Continue:* Sample from the next block if the number of cumulative deviations found in the sample falls between the values in the stop or abandon columns. These values are indicated in Table 2 of Appendix C.
- *Abandon:* If four cumulative deviations are found, the population deviation rate may be as great as the tolerable deviation rate and further sampling will not provide the desired conclusion.

In the present example, the initial block contains 44 items. If no deviations are found, *stop*. If 1-3 deviations are found, sample the second block of 40 items. If 4 deviations are found, *abandon* the sample.

20
DISCOVERY SAMPLING

Discovery sampling, a form of attribute sampling, is used when the auditor wishes to locate at least *one* deviation, given that the defined deviation conditions are occurring in the population. The sample size is designed to capture *at least one* deviation in the population. Consequently, discovery sampling is often used in fraud audits, an example of which would be looking for forged checks.

Discovery sampling can be carried out by following the rules for attribute sampling when a *one-sided upper precision limit* is desired, as discussed in Chapter 16. The only difference is that in a discovery sample, the *expected population deviation rate* is established at *zero*. If any deviations are found, such as a forged check for example, then the auditor expands the audit scope to thoroughly evaluate the extent and effect of the fraud.

REFERENCES

[10] This error could cause the sample results to support the planned assessment of control risk at less than the maximum when the population does not warrant such an assessment. The risk of assessing control risk too low is a component of *sampling risk*.

[11] In this case, the auditor should increase substantive testing *or* consider evaluating a different control procedure.

[12] A table of z values is located in Appendix E, Table 1. Locate the z value that corresponds to the desired confidence level.

[13] Use tolerable deviation rate when sample size is large; otherwise, use expected population deviation rate.

[14] Interpolation is a process to estimate the actual value when the table does not present it exactly.

[15] The sample size (n) for *one-sided upper precision limit* when no deviations are expected in the population is:

$$n = \frac{\ln(1 - \text{confidence level})}{\ln(1 - \text{tolerable deviation rate})}$$

where ln is the natural logarithm.

[16] Computed using Appendix B, Table 1.

[17] Risk of Assessing Control Risk Too Low leading to ineffectiveness.

[18] Risk of Assessing Control Risk Too High leading to inefficiency.

PART 5

PROBABILITY-PROPORTIONAL-TO-SIZE SAMPLING

21
SELECTING THE PPS SAMPLING TECHNIQUE

Probability-proportional-to-size (PPS)[19] sampling is based upon attribute sampling theory but is used for substantive audit testing. That is, PPS sampling is used to express a conclusion about the dollar values of a population. An alternative statistical sampling approach for substantive testing is **classical variables sampling**, which is presented in Part 6 of this book.

In general, PPS sampling is *easier* and *more efficient* to use than variables sampling. Variables sampling is often considered a last resort due to its technical nature and sometimes large sample sizes. Consider some of the reasons for the popularity of PPS sampling in audit practice:

- *Convenient tables* are available for the computation of PPS sample results. Variables sampling requires computer software and a statistical sampling expert.
- A PPS sample is *automatically stratified*. The larger items in a population will always be selected for the application of audit procedures by virtue of the PPS sample selection technique. The name "probability-proportional-to-size" is used because the probability of a particular item being selected is proportional to its size relative to the total. The larger the item, the higher the probability of selection; the smaller the item, the lower the probability of selection.
- When a low error rate is expected in a population, PPS sampling typically requires a much *smaller sample size* than variables sampling, leading to enhanced audit efficiency. If the auditor expects a large number of errors in the population, the PPS sample size may exceed the size required of variables sampling.

An auditor considering the use of PPS sampling should be alerted to special sample design issues that emerge in the following two situations:

1. Audited value is greater than book value on sample items (errors of understatement).
2. Zero or negative balances exist in the population.

If either condition exists, PPS sampling may not be the best choice. In situation (2), however, the auditor can consider identifying either the zero or negative balances or both and auditing them as a separate population using an approach other than PPS sampling.

22
INITIATING A PPS SAMPLING PLAN

Exhibit 22-1 summarizes the steps required to implement a PPS sampling plan. Each of the six steps must be clearly documented in the audit workpapers.

EXHIBIT 22-1
STEPS IN CONDUCTING A PPS SAMPLING PLAN

1. Determine the objectives of the audit test.
2. Define the population and the sampling unit.
3. Determine the sample size.
4. Select the sample.
5. Perform audit procedures on the sample items.
6. Evaluate the sample and express conclusion(s).

Determine the Objectives of the Audit Test

The objective of a PPS sampling plan is to express a conclusion about the population in dollar terms. In other words, with what level of confidence can the population of interest be accepted as materially correct? A PPS sampling plan is used for substantive audit testing. One of the more common applications in practice is the audit of accounts receivable. A PPS sample enables the auditor to determine whether the balance of accounts receivable is free of material error.

Define the Population and Sampling Unit

The auditor must define the *population* clearly and in advance of the sample selection. This requirement is due to the fact that the sample results can be generalized only to the population from which the sample items were selected. In the audit of accounts receivable, for example, the population may be defined as "all customer accounts with debit balances." As mentioned previously, credit and zero balances can cause difficulties in evaluating a PPS sample and should be audited separately.

The *sampling unit* in a PPS sampling plan is each *dollar* in the population. If accounts receivable has a balance of $180,000, then there are 180,000 sampling units or dollars in the population. Audit procedures are performed on the customer account that contains the *dollar* that was selected. The customer account is called a *logical unit* because it will be subjected to audit procedures. The customer account is used solely for illustration. In auditing accounts receivable, specific invoices also may be logical units. In the audit of inventory prices, the logical unit may be the value of the inventory line item. Regardless of the account or class of transactions being audited, the auditor must define *population*, *sampling unit*, and *logical unit* before determining the sample size (Chapter 23).

23
SELECTING THE SAMPLE

Determine the Sample Size

Determination of the sample size requires specification of specific key variables by the auditor. The computation is relatively simple. For ease of reference, the sample size formulas are summarized in Exhibit 23-1.

EXHIBIT 23-1
PPS SAMPLING: SAMPLE SIZE FORMULAS

$$\text{Sample Size} = \frac{\text{Population Value}}{\text{Sampling Interval *}}$$

$$\text{*Sampling Interval} = \frac{\text{Tolerable Error - (Expected Error} \cdot \text{Expansion Factor)}}{\text{Reliability Factor}}$$

The variables that must be specified by the auditor are defined below:

1. *Population Value* is the recorded book value of the population described in Chapter 22.
2. *Tolerable Error* is the auditor's assessment of materiality with respect to the population being audited. It is the amount by which the account or transaction class could be in error without causing a material misstatement in the financial statements.
3. *Expected Error* is an estimate of the dollar error that exists in the population. It should be much smaller than tolerable error. Otherwise variables sampling may be a more appropriate choice if statistical sampling is desired.
4. *Expansion Factor* comes from Appendix D, Table 2 and is dependent upon the auditor's designated risk of incorrect acceptance, or beta risk. Read across the top of the table to determine the appropriate value.
5. *Reliability Factor* is drawn from Appendix D, Table 1. When computing the *sampling interval*, the number of overstatement errors (left hand column) should be set at *zero*. The risk of incorrect acceptance is based upon the auditor's judgment.

Exhibit 23-2 lists the steps necessary to compute a PPS sample size.

EXHIBIT 23-2
STEPS TO COMPUTE A PPS SAMPLE SIZE

1. Determine the **expansion factor** (Appendix D).
2. Determine the **reliability factor** (Appendix D).
3. Compute the **sampling interval** (Exhibit 23-1).
4. Compute the **sample size** (Exhibit 23-1).

Select the Sample

PPS sampling employs *systematic sampling* which was introduced in Chapter 9. Prior to selecting the sample, however, the population must be ordered so that the cumulative dollars of the population can be identified. Here is an excerpt from an accounts receivable population:

Customer Account Number	Book Value	Cumulative Book Value	Dollar Selected
1	$ 3,250	$ 1 - 3,250	1,200
2	9,473	3,251 - 12,723	8,700
3	1,105	12,724 - 13,828	
4	6,000	13,829 - 19,828	16,200
5	1,600	19,829 - 21,428	
6	3,300	21,429 - 24,728	23,700
↓	↓	↓	↓
↓	↓	↓	↓

For the "dollar selected" column, suppose that the sampling interval is **7,500**—meaning that every **7,500th** dollar will be selected following a random start between **1** and **7,500**. If the random start is **1,200**, then the "dollars" and corresponding customer accounts identified above will be selected for the sample. Notice that the two largest accounts were selected. As mentioned previously, PPS sampling *automatically* selects large and significant items from the population.

24
EVALUATING THE RESULTS

Perform Audit Procedures on the Sample Items

This step is independent of the sampling approach used and calls for the auditor to follow the approved audit program procedures. The audited values for each logical unit should be computed and compared to the auditee's recorded values. Audit procedures and results should be clearly documented in the workpapers.

Evaluate the Sample and Express Conclusion(s)

If *no errors* are found in the sample, the audit work is complete and the auditor may conclude either of the following:

> We are (1 - risk of incorrect acceptance) confident that the book value is not overstated by more than tolerable error.
>
> *or*
>
> There is a (specified risk of incorrect acceptance) risk that the population may be overstated by more than tolerable error.

If *errors are found* in the sample, then further computations must be made by the auditor. The necessary steps are enumerated in Exhibit 24-1.

EXHIBIT 24-1
STEPS TO EVALUATE SAMPLE RESULTS IN PPS SAMPLING

1. Compute **projected error**.
2. Compute **basic precision.**
3. Compute **incremental allowance for projected error**.
4. Compute the **upper error limit**.
5. Express a conclusion.

The objective is to calculate the *upper error limit*, which is similar to the one-sided upper precision limit presented in Chapter 16. The upper error limit consists of three components:

Upper Error Limit = **Projected Error**

+ BasicPrecision

+ Incremental Allowance for Projected Error] **Allowance for Sampling Risk** (applies to Basic Precision and Incremental Allowance for Projected Error)

How to Calculate Projected Error

The easiest way to calculate projected error is to prepare a table to summarize the errors noted in the sample, such as the following (with values presented for illustration only):

(1)	(2)	(3)	(4)	(5)
Book Value	Audited Value	Tainting %	Sampling Interval	Projected Error
$ 500	$ 400	100/500 = 20%	$ 8,000	$ 1,600
900	855	45/900 = 5%	8,000	400
3,000	2,700	300/3000 = 10%	8,000	800
9,500	9,050	—[20]	NA	450
			Total Projected Error =	$ 3,250

Projected error is the auditor's best estimate of dollar error in the population based upon the sample results. It presumes that what is true about the logical unit is also true about the sampling interval. In an actual audit, the value of projected error is treated as an audit difference

Following is a description of the values in each column:

1. *Book value* is the auditee's recorded value of the logical unit sampled.
2. *Audited value* is what the auditor concludes the value should be after performing the prescribed audit procedures.
3. *Tainting %* is computed by:

$$\frac{\text{Book value - Audited value}}{\text{Book value}}$$

It measures the estimated percentage error in the sampling interval based upon the error in the logical unit. Tainting % does not apply when the value of the logical unit exceeds the value of the sampling interval. Since the entire sampling interval was sampled, error for the interval is known and need not be projected.

4. *Sampling interval* was computed to generate sample size. See Chapter 23.
5. *Projected error* is computed by multiplying the *tainting %* (3) by the *sampling interval* (4).

How to Calculate Basic Precision

Basic precision is an assumed amount of possible unknown or undiscovered error in the population due to sampling. It is similar to the upper error limit described in Chapter 16 and is computed even when no errors are found in the sample. Basic precision (**BP**) is computed as follows:

BP = Sampling Interval · Reliability Factor

The reliability factor can be found in Appendix D, Table 1.This factor should be the same used in computing the sampling interval in Chapter 23. That is, the number of overstatement errors (left hand column) should be *zero* and the risk of incorrect acceptance (across top) is defined by the auditor.

How to Calculate the Incremental Allowance for Projected Error

Only errors that were *less than* the sampling interval are considered at this stage. Preparation of a table streamlines this task. Here is an example with illustrative values:

(1) Number of Errors	(2) Reliability Factor(RF)	(3) RF Increment	(4) RF Increment - 1.00	(5) Projected Error	(6) Incremental Allowance for Projected Error
0	3.00				
1	4.75	1.75	.75	750	$ 563
2	6.30	1.55	.55	375	206

Total Incremental Allowance for Projected Error = $ 769

An analysis of each column:

1. Assume there are two overstatement errors that are less than the sampling interval, as per the computation of *projected error* presented above.
2. The reliability factor is found in Appendix D, Table 1. Locate the column with the appropriate risk of incorrect acceptance (**.05** in this example) and then identify the factors corresponding to **0**, **1**, and **2** errors.
3. Reliability factor increment is the difference in reliability factor when one more error is found. When one error is found, the increment is (4.75 - 3.00), or 1.75. When two errors are found, it is (6.30 - 4.75), or 1.55. This process continues for as many errors as are found in the sample but only when the book value with the error is less than the sampling interval.
4. The computation of incremental allowance for projected error uses the reliability factor increment minus 1.00. Preparation of this column facilitates the computation.
5. Projected error comes directly from the previous calculation of "projected error." In computing the incremental allowance for projected error, however, the projected errors are *ranked from largest to smallest*. This approach results in more emphasis being placed on the larger errors, leading to a more conservative analysis.
6. The incremental allowance for projected error is computed by multiplying the value in column (**4**) by the value in column (**5**). All values in column (**6**) are summed to arrive at total incremental allowance for projected error.

Concluding from the Sample Results

Suppose the following amounts were obtained with a **.05** risk of incorrect acceptance:

Projected Error	=	$ 9,250
+ Basic Precision	=	13,600
+ Incremental Allowance for Projected Error	=	1,061
Upper Error Limit		**$23,911**

If the upper error limit is less than tolerable error, then the recorded book value can be accepted as materially correct and the auditor can express the following conclusions:

> There is a 5 percent risk that the population is overstated by more than tolerable error.
>
> *or*
>
> We are 95 percent confident that the population is not overstated by more than tolerable error.

If the upper error limit is *greater than* tolerable error, then the recorded book value may be materially overstated. If this situation occurs, the auditor must expand the scope of the audit to pinpoint more closely the value and nature of the overstatement errors. Some suggestions include:

1. Examine more logical units from the population.
2. Perform other substantive audit procedures.
3. Have the auditee record a journal entry for the actual errors found and *reduce* the *upper error limit* accordingly. Compare the revised value to tolerable error to determine if the population can be accepted.

REFERENCES

[19] Also known by the following terms:

- Dollar unit sampling (DUS).
- Combined attributes variables (CAV) sampling.
- Cumulative monetary amount sampling (CMA).
- Monetary unit sampling (MUS).

[20] When the logical unit (recorded book value) exceeds the sampling interval, the *actual error* in the logical unit is carried to the "projected error" column.

PART 6

CLASSICAL VARIABLES SAMPLING

25
TYPES OF CLASSICAL VARIABLES SAMPLING

Classical variables sampling is used for substantive tests of details. This technique is very complex and nearly impossible to perform manually. Consequently, auditors rely on computer software to perform the calculations.

Variables sampling is based on normal distribution theory. This theory permits the auditor to make inferences about the population by examining only a portion of it. The auditor need not be concerned with the *distribution* of the population as a result of the *central limit theorem*, which states:

> The distribution of the sample mean tends to become normal as the sample size increases. When samples are unbiased, the mean of the sampling distribution is equal to the population mean.

This theorem is more understandable after the terms are explained.

Suppose all possible samples of size **n** are drawn from a population and the mean of each sample is calculated. The *distribution* of the sample means will be approximately *normal*, provided the sample size is 30 or more. Further, the *mean* of the sampling distribution will equal the true population mean.

The sampling distribution is a theoretical concept, since it is cost and time prohibitive for auditors to take repeated samples from the same population. In practice, only *one* sample is drawn to make estimates about the true population value. Since only one sample is used, the resulting estimate of the true population value probably will not be exact. Instead, *precision intervals*[21] are constructed around the estimate of the mean. Some statistics useful for this task are presented in Exhibit 25-1. Refer to Part 2 for a refresher on the computation of these.

EXHIBIT 25-1
STATISTICS USED IN VARIABLES SAMPLING

	Mean	Standard Deviation
Sample	$\bar{x}$	s
Population	μ	σ
Sampling Distribution	$\mu_{\bar{x}}$	$\sigma_{\bar{x}}$*

* Called **standard error of the mean.**

The standard error of the mean can be expressed as:

$$\sigma_{\bar{x}} = \frac{\sigma}{\sqrt{n}}$$

This relationship reveals that $\sigma_{\bar{x}}$ can be made smaller by increasing the sample size, **n**. The smaller that $\sigma_{\bar{x}}$ becomes, the more precise the estimate of the population mean will be.

The sampling distribution of the sample mean can be represented by the following diagram:

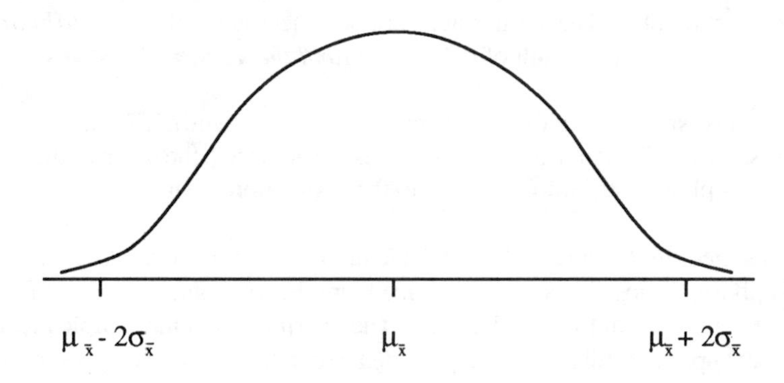

Suppose that $\sigma_{\bar{x}}$ = \$75 and $\mu_{\bar{x}}$ = \$26,250. Then the following equivalent statements can be made about the population:

- 95.44 percent[22] of all sample means will lie within two standard errors of the population mean.
- There exists a 95.44 percent probability that *any* sample mean will lie within two standard errors of the mean.
- We are 95.44 percent confident that the true population mean is within the interval:

$$(1) \quad \bar{x} \pm 2\sigma_{\bar{x}}$$

The confidence can be expressed in terms of a *precision interval*. The auditor can vary the size of the precision interval by substituting other values for 2 in formula (1) above. These "other values" are called *z values* and measure distance from the mean in terms $\sigma_{\bar{x}}$. Appendix E, Table 1 contains various *z values* and the corresponding confidence levels that are useful in variables sampling.

The previous discussion demonstrates that normal distribution theory, particularly the central limit theorem, enables auditors to estimate true population values. Since the estimate is based on sampling, the exact value of the population cannot be known. However, a precision interval can be constructed to define a range of reasonable values at a specified level of confidence. The *z values* in Appendix E, Table 1 can be used to construct precision intervals of various sizes, although most audit applications use *at least* 90 percent confidence.

26
THREE APPROACHES TO VARIABLES SAMPLING

There are three approaches to variables sampling in auditing. Selection of an approach depends upon the data being audited. The choices are:

- Mean per unit estimation.
- Difference estimation.
- Ratio estimation.

Exhibit 26-1 summarizes the conditions where each is best suited. Many computer programs are designed to select the approach that provides the most efficient sample size. An example of how each approach works is presented in Exhibit 26-2.

EXHIBIT 26-1
CONDITIONS LEADING TO SELECTION OF VARIABLES SAMPLING APPROACH

	Mean Per Unit Estimation	Ratio Estimation	Difference Estimation
• Book value of population unknown	♦		
• Book value of population known		♦	♦
• Sample expected to include at least 50 errors		♦	♦
• Population can be stratified	♦	♦	♦
• Variability of book values is:			
High		♦	♦
Low	♦	♦	♦
• Magnitude of errors expected to be similar			♦
• Ratio of audited values to book values similar		♦	

EXHIBIT 26-2
THREE APPROACHES TO VARIABLES SAMPLING

- *Mean per unit estimation*

Formula:

Average audited value of sample items	·	Number of items in population	=	Estimated population value

- *Ratio estimation*

Formula:

Sum of audited values in sample / Sum of book values in sample	·	Book value of population	=	Estimated population value

- *Difference estimation*

Formula:

Average difference between audited and book values in sample	·	Number of items in population	=	Estimated **difference** (between audited and book values in population)

27
INITIATING A VARIABLES SAMPLING PLAN

Exhibit 27-1 outlines the six steps the auditor should carry out in a variables sampling plan. Each step should be clearly documented in the audit workpapers. The complexity of variables sampling requires that the documentation be especially detailed.

EXHIBIT 27-1
STEPS IN CARRYING OUT A VARIABLES SAMPLING PLAN

1. Determine the objectives of the audit test.
2. Define the population and the sampling unit.
3. Determine the sample size.
4. Select the sample.
5. Perform audit procedures on the sample items.
6. Evaluate the sample and express conclusion(s).

Determine the Objectives of the Audit Test

The objective of a variables sampling plan is to reach a conclusion about the dollar value of the population being audited. Therefore it is used for substantive testing of details. Most audit applications are designed to test the *hypothesis* "the account is not misstated by more than a material amount (tolerable error)." However, variables sampling is also useful when the auditor wishes to independently *estimate* the unknown value of a population. An example of using variables sampling to estimate is a conversion from FIFO to LIFO—management may desire an estimate of the LIFO value of the FIFO-based inventory.

Define the Population and Sampling Unit

The auditor must formally define the population and sampling unit. The population is the account balance or class of transactions being audited. The sampling unit is the element within the population that will be subjected to audit procedures. Some examples are presented in Exhibit 27-2. Sample results can be projected only to the defined population from which the sample is drawn.

EXHIBIT 27-2
EXAMPLES OF HOW TO DEFINE THE POPULATION AND SAMPLING UNIT

Population	Sampling Unit
• Accounts Receivable	• Customer Account or Invoice
• Inventories	• Inventory Item
• Loans Receivable	• Customer Account
• Sales	• Invoice

Additional considerations for the auditor in defining the population include:

- Ensure that the sample frame is a *complete* representation of the population. All items must have a known probability of selection for statistical procedures to be valid. Always test the sample frame for completeness before selecting the sample.
- *Individually significant items* are those that could individually exceed or equal materiality (tolerable error). These items should be removed from the population and audited separately in their entirety. That is, the auditor should examine 100 percent of the individually significant items as a separate population.

28
DETERMINING THE SAMPLE SIZE

Six factors impact sample size in variables sampling and are summarized in Exhibit 28-1. Auditors should be familiar with each factor and its relationship to sample size because the auditor can control the sample size by altering the factors. Notice the similarity between the variables sampling factors and the attribute sampling factors that were presented in Exhibit 13.1.

EXHIBIT 28-1
FACTORS THAT IMPACT SAMPLE SIZE IN VARIABLES SAMPLING

Factor	Impact on Sample Size
• Population Size	• Direct
• Risk of Incorrect Acceptance (β risk)	• Inverse
• Risk of Incorrect Rejection (α risk)	• Inverse
• Tolerable Error (precision)	• Inverse
• Expected Error	• Direct
• Variation in population	• Direct

Direct = as sample size factor increases (decreases), sample size becomes larger (smaller).

Inverse = as sample size factor increases (decreases), sample size becomes smaller (larger).

Population size has a direct relationship to sample size. Generally speaking, the larger the population, the larger the sample size.

The *risk of incorrect acceptance* is the component of sampling risk that measures the risk that the auditor will accept a book value as being correct when in fact it contains a material error. This risk is also called beta (β) risk and results in audit *ineffectiveness.* Beta risk is typically established in the neighborhood of five percent due to the serious consequences of committing this type of error. However, its value is a function of the auditor's judgment coupled with policies established by the internal auditing department. Beta risk bears an *inverse* relationship to sample size because the more risk the auditor is willing to incur, the smaller the number of items that will need to be audited.

The *risk of incorrect rejection*, or alpha (α) risk, is the risk leading to audit *inefficiency.* Alpha risk also has an *inverse* relationship to sample size for the same reasons as beta risk. However, since alpha risk does not lead to ineffectiveness, the auditor may feel justified in establishing it at a higher level than beta risk.

Tolerable error is the auditor's assessment of materiality for the population. It represents the error that the auditor can tolerate and still accept the population or recorded book value as being materially correct. The greater the tolerable error, the smaller the required sample size. Tolerable error, therefore, bears an *inverse* relationship to sample size.

Expected error is an estimate of how much dollar error exists in the population. The auditor computes expected error based upon experience, familiarity with the business, inquiry, or a pilot sample. In a variables sampling plan, expected error should be greater than zero since errors are required in the sample to yield meaningful results in ratio and difference estimation. The relationship between expected error and sample size is *direct*—as expected error increases, so does the required sample size.

Variation is measured by σ . The greater the variation, the greater the sample size needed to capture a representative selection of items. Many auditors *stratify* (see Chapter 10) to reduce the effect of variation on sample size. Some computer programs will stratify automatically.

Sample Size Formula

$$n = \left(\frac{N \cdot \sigma \cdot z^*}{A} \right)^2$$

Where: n = sample size

N = number of items in the population

s = estimated population standard deviation

z* = z value corresponding to desired level of confidence (see Appendix E, Table 1)

A = planned precision[23]

An illustration of the *mean per unit* approach to variables sampling will use the following hypothetical data relating to the audit of accounts receivable:

Population	9,500 customer accounts
Book value	$1,911,000
Estimated	$26
Tolerable error	$97,500
Beta risk	.05
Alpha risk	.05
Confidence level	95%

Now substitute these values into the sample size formula:

$$n = \left(\frac{19{,}500 \cdot 26 \cdot 1.96^{24}}{97{,}500 \cdot .543} \right)^2$$

$$n = \left(\frac{993{,}720}{52{,}943} \right)^2$$

$$n = (18.77)^2 \approx \mathbf{352}$$

This computed sample size can be reduced further with the *finite population correction factor* (fpc) because the above formula assumes an infinite population—untrue in audit applications. The fpc is combined with the above formula to yield the formula for the *actual sample size*, $\mathbf{n_a}$:

$$n_a = \frac{n}{1 + n/N}$$

Using the data presented earlier:

$$n_a = \frac{352}{1 + 352/19{,}500}$$

$$n_a = \frac{352}{1.018} = \mathbf{346}$$

Notice that the fpc has a minimal impact in this example—reducing the sample size by only three items. As a practical matter, n_a is used only when the sample size is greater than or equal to 20 percent of the population since the effect is otherwise insignificant.

Select the Sample

Since variables sampling is a statistical technique, a random selection method must be used by the auditor. The most frequently employed methods are simple random sampling and systematic selection. Refer to Chapter 9 for details on how to select samples for each of these.

29
EVALUATING THE RESULTS

Perform the Audit Procedures on the Sample Items

Performance of the audit procedures is independent of the sampling approach used. At this stage, the auditor performs the required audit procedures according to the approved audit program. The working papers should contain a summary of the book value and audited value for each item in the sample, as these will be required to evaluate the sample and generalize results into the population.

Evaluate the Sample Results and Express Conclusion(s)

The manner in which the sample results are evaluated depends upon which of the three variables sampling approaches are used—mean per unit, ratio, or difference estimation. Refer to Exhibit 26-2 for how to compute each. Given the hypothetical data presented in Chapter 28, suppose the sample revealed the following results (recall this is a mean per unit application):

Average Audited Value of Items in Sample	$130
Sample Standard Deviation	$ 25

The projected value of accounts receivable based on the sample results is:

Average audited value of sample items	·	Number of items in population	=	Projected population value
$130	·	19,500	=	$2,535,000

Now a precision interval must be constructed around the projected population value. This procedure is a two-step process that is the same regardless of which approach is used (mean per unit, ratio, or difference estimation). The necessary formulas are presented in Exhibit 29-1.

EXHIBIT 29-1
FORMULAS FOR EVALUATION OF RESULTS FROM A VARIABLES SAMPLE

Achieved Precision

$$A_a = N \cdot z^* \cdot \frac{s\sqrt{1 - (n_a \div N)}}{\sqrt{n_a}}$$

Where:
- A_a = achieved precision
- N = population size
- z^* = *z value* corresponding to desired confidence level (see Appendix E, Table 1)
- s = sample standard deviation
- n_a = sample size adjusted for finite population correction factor

Adjusted Achieved Precision

$$A_j = A_a + TE\,[1 - (A_a \div A)]$$

Where:
- A_j = adjusted achieved precision
- A_a = achieved precision
- TE = tolerable error
- A = planned precision

How to Interpret Output from Variables Sampling Programs

An example of output from a variables sampling program is presented in Exhibit 29-2. The items specified by the auditor are underlined. The values produced by the program are not. The output can be interpreted exactly as the manual computations were interpreted with an additional advantage—the program will print out the conclusion, leaving no doubt in the auditor's mind. For example, if the sample results indicated that there was an error that exceeded tolerable error in the book value of the account being audited, it would print out a message similar to the following:

> THE DESIRED PRECISION AT THE SPECIFIED CONFIDENCE LEVEL WAS NOT ACHIEVED IN THIS SAMPLE!! ADDITIONAL SAMPLING IS REQUIRED.

The output will also indicate the actual confidence that the auditor can place in the results. In the case of a sample that did not achieve the desired confidence, the output might read:

> YOU CAN BE 91% CONFIDENT THAT THE POPULATION VALUE IS WITHIN THE SPECIFIED RANGE OF TOLERABLE ERROR.

The auditor then can evaluate whether 91 percent is sufficient to draw a conclusion or whether additional sampling will be required.

EXHIBIT 29-2
EXAMPLE VARIABLES SAMPLING OUTPUT

NAME OF AUDITEE: Subsidiary RBG

ACCOUNT SAMPLED: Accounts Receivable

APPROACH USED: Mean Per Unit Estimation

SPECIFICATIONS OF SAMPLE SIZE FACTORS:

TOLERABLE ERROR	$75,000
BETA RISK	.05
ALPHA RISK	.05
CONFIDENCE LEVEL	.95
BOOK VALUE	$1,470,000
NO. ITEMS IN POPULATION	15,000
REQUIRED SAMPLE SIZE	206

SAMPLE EVALUATION:

PROJECTED VALUE OF ACCOUNT	$1,500,000
UPPER PRECISION LIMIT	$1,542,485
LOWER PRECISION LIMIT	$1,457,515

ANY VALUE IN THE RANGE OF $1,457,515 TO $1,542,485 CAN BE ACCEPTED WITH AT LEAST 95 PERCENT CONFIDENCE THAT IT IS NOT IN ERROR BY AN AMOUNT THAT EXCEEDS TOLERABLE ERROR.

REFERENCES

[21] Also known as confidence intervals.

[22] Per Appendix E, Table 1, this is the confidence level associated with a distance of two standard deviations from the mean.

[23] **A** = planned precision. This is computed by multiplying **tolerable error** by the value from Appendix E, Table 2 that corresponds to the specified alpha and beta risks.

[24] See Appendix E, Table 1 for the *z value* corresponding to 95 percent confidence.

APPENDICES

APPENDIX A
TABLES FOR COMPUTING SAMPLE SIZE AND TWO-SIDED PRECISION INTERVALS FORATTRIBUTE SAMPLING

Table 1: Statistical Sample Sizes for Attribute Sampling
(Two-Sided Precision Interval)
90% Confidence Level

Table 2: Statistical Sample Sizes for Attribute Sampling
(Two-Sided Precision Interval)
95% Confidence Level

Table 3: Statistical Sample Evaluation for Attribute Sampling
(Two-Sided Precision Intervals)
Sample Deviation Rate = 0%

Table 4: Statistical Sample Evaluation for Attribute Sampling
(Two-Sided Precision Intervals)
Sample Deviation Rate = 2%

Table 5: Statistical Sample Evaluation for Attribute Sampling
(Two-Sided Precision Intervals)
Sample Deviation Rate = 3%

Table 6: Statistical Sample Evaluation for Attribute Sampling
(Two-Sided Precision Intervals)
Sample Deviation Rate = 5%

Table 7: Statistical Sample Evaluation for Attribute Sampling
(Two-Sided Precision Intervals)
Sample Deviation Rate = 10%

These tables were adapted from *Sampling for Modern Auditors* published by The Institute of Internal Auditors, Inc., Altamonte Springs, Florida, 1977.

Table 1

Statistical Sample Sizes for Attribute Sampling (Two-Sided Precision Interval) 90% Confidence Level

a. Expected Population Deviation Rate Not Over 2%

Precision Percentage +/-

Population	.50	.75	1.00	1.25	1.50	1.75	2.00
1,000	679	485	346	253	190	147	117
2,000	1029	640	419	290	210	159	124
5,000	1489	793	479	317	225	167	129
10,000	1750	861	503	328	230	170	130
50,000 +	2035	925	524	337	234	172	132

b. Expected Population Deviation Rate Not Over 3%

Precision Percentage +/-

Population	.50	.75	1.00	1.25	1.50	1.75	2.00
1,000	759	440	259	164	111	94	80
2,000	1223	564	297	179	118	98	83
5,000	1932	680	327	189	122	102	85
10,000	2395	729	338	193	124	103	86
50,000 +	2963	775	347	196	125	103	87

c. Expected Population Deviation Rate Not Over 5%

Precision Percentage +/-

Population	.50	.75	1.00	1.25	1.50	1.75	2.00
1,000	562	363	243	170	124	94	74
2,000	782	444	276	186	133	99	77
5,000	1022	512	301	197	138	102	79
10,000	1138	540	311	201	140	103	79
50,000 +	1253	564	319	204	142	104	80

Table 2

Statistical Sample Sizes for Attribute Sampling (Two-Sided Precision Interval) 95% Confidence Level

a. Expected Population Deviation Rate Not Over 2%

Precision Percentage +/-

Population	.50	.75	1.00	1.25	1.50	1.75	2.00
1,000	750	572	429	325	250	197	158
2,000	1201	801	547	388	286	218	172
5,000	1879	1055	654	439	313	234	181
10,000	2314	1180	700	459	323	239	184
50,000 +	2840	1303	741	477	332	244	187

b. Expected Population Deviation Rate Not Over 3%

Precision Percentage +/-

Population	.50	.75	1.00	1.25	1.50	1.75	2.00
1,000	817	527	331	218	151	128	110
2,000	1381	717	397	245	164	137	116
5,000	2360	913	451	264	172	143	121
10,000	3089	1005	473	271	175	145	122
50,000 +	4104	1093	491	277	178	147	123

c. Expected Population Deviation Rate Not Over 5%

Precision Percentage +/-

Population	.50	.75	1.00	1.25	1.50	1.75	2.00
1,000	645	447	313	225	168	129	102
2,000	954	577	371	254	184	138	107
5,000	1336	697	418	275	194	144	111
10,000	1543	750	436	283	198	146	112
50,000 +	1760	798	452	290	201	148	113

Table 3

Statistical Sample Evaluation for Attribute Sampling (Two-Sided Precision Intervals) Based Upon Deviation Rate Found in Sample

Sample Deviation Rate = 0%

Sample Size	Population 1,000 UL	2,000 UL	10,000 UL	50,000+ UL
	90% Confidence Level			
30	7.3	7.3	7.4	7.4
40	5.5	5.5	5.6	5.6
50	4.4	4.4	4.5	4.5
60	3.7	3.7	3.8	3.8
70	3.1	3.2	3.2	3.2
80	2.7	2.8	2.8	2.8
90	2.4	2.5	2.5	2.5
100	2.2	2.2	2.3	2.3
150	1.4	1.5	1.5	1.5
200	1.0	1.1	1.1	1.1
300	.6	.7	.7	.8
400	.4	.5	.6	.6
500	.3	.4	.5	.5
1,000		.2	.2	.2
	95% Confidence Level			
30	9.4	9.4	9.5	9.5
40	7.1	7.1	7.2	7.2
50	5.7	5.7	5.8	5.8
60	4.7	4.8	4.9	4.9
70	4.0	4.1	4.2	4.2
80	3.5	3.6	3.7	3.7
90	3.1	3.2	3.3	3.3
100	2.8	2.9	2.9	3.0
150	1.8	1.9	2.0	2.0
200	1.3	1.4	1.5	1.5
300	.8	.9	1.0	1.0
400	.6	.7	.7	.7
500		.5	.6	.6
1,000			.3	.3

UL = Upper Precision Limit
NOTE: Lower Precision Limit = 0 when sample deviation rate = 0%

Table 4

Statistical Sample Evaluation for Attribute Sampling (Two-Sided Precision Intervals) Based Upon Deviation Rate Found in Sample

Sample Deviation Rate = 2%

	Population							
	1,000		2,000		10,000		50,000+	
Sample Size	LL	UL	LL	UL	LL	UL	LL	UL
	90% Confidence Level							
50	.2	9.1	.1	9.2	.1	9.3	.1	9.3
80	.4	6.8	.3	6.9	.3	7.0	.3	7.0
90	.4	6.4	.4	6.5	.3	6.6	.3	6.6
100	.4	6.0	.4	6.1	.4	6.2	.4	6.2
120	.6	5.5	.5	5.6	.5	5.7	.5	5.7
140	.6	5.1	.6	5.2	.5	5.3	.5	5.3
150	.7	4.9	.6	5.0	.6	5.1	.6	5.1
180	.8	4.6	.7	4.7	.7	4.8	.7	4.8
200	.8	4.3	.8	4.4	.7	4.5	.7	4.5
250	1.0	4.0	.9	4.2	.9	4.3	.9	4.3
300	1.2	3.5	1.1	3.7	1.0	3.8	1.0	3.8
500	1.4	2.9	1.3	3.2	1.2	3.3	1.2	3.3
700			1.4	2.9	1.3	3.0	1.3	3.1
1,000			1.6	2.6	1.4	2.8	1.4	2.8
	95% Confidence Level							
50	.1	10.4	.1	10.6	.1	10.6	.1	10.6
80	.3	7.7	.2	7.8	.2	7.9	.2	8.0
90	.3	7.2	.3	7.3	.2	7.4	.2	7.4
100	.3	6.8	.3	6.9	.3	7.0	.2	7.0
120	.4	6.1	.4	6.3	.3	6.4	.3	6.4
140	.5	5.7	.4	5.8	.4	6.0	.4	6.0
150	.5	5.4	.5	5.6	.4	5.7	.4	5.7
180	.7	5.1	.6	5.2	.5	5.3	.5	5.4
200	.7	4.7	.6	4.9	.6	5.0	.6	5.0
250	.9	4.4	.8	4.6	.7	4.8	.7	4.8
300	1.1	3.8	1.0	4.0	.9	4.2	.9	4.2
500			1.2	3.4	1.1	3.5	1.1	3.6
700			1.3	3.0	1.2	3.2	1.2	3.3
1,000					1.3	3.0	1.3	3.0

LL = Lower Precision Limit
UL = Upper Precision Limit

Table 5

Statistical Sample Evaluation for Attribute Sampling (Two-Sided Precision Intervals) Based Upon Deviation Rate Found in Sample

Sample Deviation Rate = 3%

	Population							
	1,000		2,000		10,000		50,000+	
Sample Size	LL	UL	LL	UL	LL	UL	LL	UL
	90% Confidence Level							
80	.8	8.2	.7	8.3	.7	8.4	.7	8.4
90	.9	7.8	.8	7.9	.8	8.0	.8	8.0
100	.9	7.4	.9	7.5	.8	7.6	.8	7.6
120	1.1	6.8	1.0	6.9	1.0	7.0	1.0	7.0
140	1.2	6.4	1.2	6.6	1.1	6.7	1.1	6.7
150	1.3	6.2	1.2	6.4	1.1	6.5	1.1	6.5
180	1.4	5.9	1.4	6.0	1.3	6.2	1.3	6.2
200	1.5	5.6	1.4	5.7	1.4	5.8	1.3	5.9
250	1.7	5.3	1.6	5.5	1.6	5.6	1.5	5.7
300	1.9	4.8	1.8	4.9	1.7	5.1	1.7	5.1
500	2.3	4.1	2.1	4.3	2.0	4.5	2.0	4.5
700			2.3	4.0	2.2	4.2	2.1	4.2
1,000			2.5	3.7	2.3	3.9	2.3	4.0
	95% Confidence Level							
80	.6	9.2	.6	9.3	.5	9.5	.5	9.5
90	.7	8.7	.6	8.8	.6	8.9	.6	8.9
100	.7	8.3	.7	8.4	.6	8.5	.6	8.5
120	.9	7.5	.8	7.6	.8	7.8	.8	7.8
140	1.0	7.1	.9	7.3	.9	7.4	.9	7.4
150	1.1	6.8	1.0	7.0	.9	7.1	.9	7.1
180	1.2	6.4	1.1	6.6	1.1	6.8	1.1	6.8
200	1.3	6.1	1.2	6.2	1.1	6.4	1.1	6.4
250	1.5	5.8	1.4	6.0	1.3	6.1	1.3	6.2
300	1.8	5.1	1.6	5.3	1.5	5.5	1.5	5.5
500			2.0	4.6	1.8	4.8	1.8	4.8
700			2.1	4.2	2.0	4.4	2.0	4.5
1,000					2.2	4.1	2.1	4.2

LL = Lower Precision Limit
UL = Upper Precision Limit

Table 6

Statistical Sample Evaluation for Attribute Sampling (Two-Sided Precision Intervals) Based Upon Deviation Rate Found in Sample

Sample Deviation Rate = 5%

	Population							
	1,000		2,000		10,000		50,000+	
Sample Size	LL	UL	LL	UL	LL	UL	LL	UL
	90% Confidence Level							
80	1.8	10.9	1.8	11.0	1.7	11.1	1.7	11.1
90	2.0	10.4	1.9	10.5	1.9	10.6	1.9	10.7
100	2.1	10.0	2.1	10.1	2.0	10.3	2.0	10.3
120	2.4	9.4	2.3	9.5	2.2	9.6	2.2	9.7
140	2.6	9.0	2.5	9.2	2.4	9.3	2.4	9.3
150	2.7	8.7	2.6	8.9	2.5	9.0	2.5	9.0
180	2.9	8.3	2.8	8.4	2.7	8.6	2.7	8.6
200	3.0	8.0	2.9	8.2	2.8	8.3	2.8	8.4
250	3.3	7.5	3.1	7.7	3.0	7.9	3.0	7.9
300	3.4	7.2	3.3	7.4	3.2	7.6	3.2	7.6
500	4.0	6.4	3.7	6.7	3.6	6.9	3.5	6.9
700			4.0	6.3	3.8	6.5	3.8	6.6
1,000			4.2	5.9	4.0	6.2	3.9	6.3
	95% Confidence Level							
80	1.5	12.0	1.5	12.2	1.4	12.3	1.4	12.3
90	1.7	11.4	1.6	11.6	1.5	11.7	1.5	11.7
100	1.8	11.0	1.7	11.1	1.7	11.3	1.6	11.3
120	2.1	10.2	2.0	10.4	1.9	10.5	1.9	10.6
140	2.3	9.8	2.2	10.0	2.1	10.1	2.1	10.1
150	2.3	9.4	2.2	9.6	2.1	9.8	2.1	9.8
180	2.6	8.9	2.4	9.1	2.3	9.2	2.3	9.3
200	2.7	8.6	2.6	8.8	2.5	9.0	2.4	9.0
250	3.0	8.0	2.8	8.3	2.7	8.4	2.7	8.5
300	3.2	7.6	3.0	7.9	2.9	8.1	2.8	8.1
500			3.5	7.0	3.3	7.2	3.3	7.3
700			3.8	6.5	3.6	6.8	3.5	6.9
1,000					3.8	6.5	3.8	6.5

LL = Lower Precision Limit
UL = Upper Precision Limit

Table 7

Statistical Sample Evaluation for Attribute Sampling (Two-Sided Precision Intervals) Based Upon Deviation Rate Found in Sample

Sample Deviation Rate = 10%

	Population							
	1,000		2,000		10,000		50,000+	
Sample Size	LL	UL	LL	UL	LL	UL	LL	UL
	90% Confidence Level							
50	4.2	19.7	4.1	19.8	4.0	19.8	4.0	19.9
80	5.3	17.1	5.2	17.2	5.1	17.3	5.1	17.3
90	5.6	16.5	5.5	16.7	5.4	16.8	5.4	16.8
100	5.8	16.1	5.6	16.2	5.6	16.4	5.5	16.4
120	6.1	15.4	6.0	15.5	5.9	15.7	5.9	15.7
140	6.4	14.8	6.3	15.0	6.2	15.2	6.2	15.2
150	6.6	14.4	6.4	14.6	6.3	14.8	6.3	14.8
180	6.9	14.1	6.7	14.3	6.6	14.4	6.6	14.5
200	7.1	13.8	6.9	14.0	6.8	14.2	6.8	14.2
250	7.5	13.2	7.3	13.5	7.1	13.7	7.1	13.7
300	7.8	12.8	7.5	13.1	7.4	13.3	7.3	13.3
500	8.5	11.8	8.2	12.2	8.0	12.4	7.9	12.5
700			8.6	11.7	8.3	12.0	8.2	12.1
1,000			8.9	11.2	8.6	11.6	8.5	11.7
	95% Confidence Level							
50	3.5	21.5	3.4	21.7	3.3	21.8	3.3	21.8
80	4.6	18.4	4.5	18.6	4.4	18.7	4.4	18.8
90	4.9	17.8	4.8	17.9	4.7	18.1	4.7	18.1
100	5.2	17.3	5.0	17.4	4.9	17.6	4.9	17.6
120	5.6	16.4	5.4	16.6	5.3	16.8	5.3	16.8
140	5.9	15.8	5.7	16.0	5.6	16.2	5.6	16.2
150	6.0	15.5	5.9	15.7	5.7	15.9	5.7	15.9
180	6.4	14.8	6.2	15.1	6.1	15.3	6.0	15.3
200	6.6	14.5	6.4	14.8	6.3	15.0	6.2	15.0
250	7.0	13.8	6.8	14.1	6.6	14.4	6.6	14.4
300	7.4	13.3	7.1	13.7	6.9	13.9	6.9	14.0
500			7.9	12.6	7.6	12.9	7.5	13.0
700			8.3	12.0	8.0	12.4	7.9	12.5
1,000					8.3	11.9	8.2	12.0

LL = Lower Precision Limit
UL = Upper Precision Limit

APPENDIX B
TABLES FOR COMPUTING SAMPLE SIZE AND ONE-SIDED UPPER PRECISION LIMITS FOR ATTRIBUTE SAMPLING

Table 1: Statistical Sample Sizes for Attribute Sampling
Five-Percent (5%) Risk of Assessing Control Risk Too Low

Table 2: Statistical Sample Sizes for Attribute Sampling
Ten-Percent (10%) Risk of Assessing Control Risk Too Low

Table 3: Statistical Sample Results Evaluation Table for Attribute Sampling
Upper Limits at Five-Percent Risk of Assessing Control Risk Too Low

Table 4: Statistical Sample Results Evaluation Table for Attribute Sampling
Upper Limits at Ten-Percent Risk of Assessing Control Risk Too Low

Table 1

Statistical Sample Sizes for Attribute Sampling Five-Percent (5%) Risk of Assessing Control Risk Too Low (with number of expected deviations in parentheses)

Expected Population Deviation Rate	Tolerable Deviation Rate										
	2%	3%	4%	5%	6%	7%	8%	9%	10%	15%	20%
0.00%	149(0)	99(0)	74(0)	59(0)	49(0)	42(0)	36(0)	32(0)	29(0)	19(0)	14(0)
0.25	236(1)	157(1)	117(1)	93(1)	78(1)	66(1)	58(1)	51(1)	46(1)	30(1)	22(1)
0.50	*	157(1)	117(1)	93(1)	78(1)	66(1)	58(1)	51(1)	46(1)	30(1)	22(1)
0.75	*	208(2)	117(1)	93(1)	78(1)	66(1)	58(1)	51(1)	46(1)	30(1)	22(1)
1.00	*	*	156(2)	93(1)	78(1)	66(1)	58(1)	51(1)	46(1)	30(1)	22(1)
1.25	*	*	156(2)	124(2)	78(1)	66(1)	58(1)	51(1)	46(1)	30(1)	22(1)
1.50	*	*	192(3)	124(2)	103(2)	66(1)	58(1)	51(1)	46(1)	30(1)	22(1)
1.75	*	*	227(4)	153(3)	103(2)	88(2)	77(2)	51(1)	46(1)	30(1)	22(1)
2.00	*	*	*	181(4)	127(3)	88(2)	77(2)	68(2)	46(1)	30(1)	22(1)
2.25	*	*	*	208 (5)	127(3)	88(2)	77(2)	68(2)	61(2)	30(1)	22(1)
2.50	*	*	*	*	150(4)	109(3)	77(2)	68(2)	61(2)	30(1)	22(1)
2.75	*	*	*	*	173(5)	109(3)	95(3)	68(2)	61(2)	30(1)	22(1)
3.00	*	*	*	*	195(6)	129(4)	95(3)	84(3)	61(2)	30(1)	22(1)
3.25	*	*	*	*	*	148(5)	112(4)	84(3)	61(2)	30(1)	22(1)
3.50	*	*	*	*	*	167(6)	112(4)	84(3)	76(3)	40(2)	22(1)
3.75	*	*	*	*	*	185(7)	129(5)	100(4)	76(3)	40(2)	22(1)
4.00	*	*	*	*	*	*	146(6)	100(4)	89(4)	40(2)	22(1)
5.00	*	*	*	*	*	*	*	158(8)	116(6)	40(2)	30(2)
6.00	*	*	*	*	*	*	*	*	179(11)	50(3)	30(2)
7.00	*	*	*	*	*	*	*	*	*	68(5)	37(3)

*Sample size is too large to be cost-effective for most audit applications.

NOTE: This table assumes a large population.

Table 2

Statistical Sample Sizes for Attribute Sampling Ten-Percent (10%) Risk of Assessing Control Risk Too Low (with number of expected deviations in parentheses)

Expected Population Deviation Rate	Tolerable Deviation Rate										
	2%	3%	4%	5%	6%	7%	8%	9%	10%	15%	20%
0.00%	114(0)	76(0)	57(0)	45(0)	38(0)	32(0)	28(0)	25(0)	22(0)	15(0)	11(0)
0.25	194(1)	129(1)	96(1)	77(1)	64(1)	55(1)	48(1)	42(1)	38(1)	25(1)	18(1)
0.50	194(1)	129(1)	96(1)	77(1)	64(1)	55(1)	48(1)	42(1)	38(1)	25(1)	18(1)
0.75	265(2)	129(1)	96(1)	77(1)	64(1)	55(1)	48(1)	42(1)	38(1)	25(1)	18(1)
1.00	*	176(2)	96(1)	77(1)	64(1)	55(1)	48(1)	42(1)	38(1)	25(1)	18(1)
1.25	*	221(3)	132(2)	77(1)	64(1)	55(1)	48(1)	42(1)	38(1)	25(1)	18(1)
1.50	*	*	132(2)	105(2)	64(1)	55(1)	48(1)	42(1)	38(1)	25(1)	18(1)
1.75	*	*	166(3)	105(2)	88(2)	55(1)	48(1)	42(1)	38(1)	25(1)	18(1)
2.00	*	*	198(4)	132(3)	88(2)	75(2)	48(1)	42(1)	38(1)	25(1)	18(1)
2.25	*	*	*	132(3)	88(2)	75(2)	65(2)	42(1)	38(1)	25(1)	18(1)
2.50	*	*	*	158(4)	110(3)	75(2)	65(2)	58(2)	38(1)	25(1)	18(1)
2.75	*	*	*	209(6)	132(4)	94(3)	65(2)	58(2)	52(2)	25(1)	18(1)
3.00	*	*	*	*	132(4)	94(3)	65(2)	58(2)	52(2)	25(1)	18(1)
3.25	*	*	*	*	153(5)	113(4)	82(3)	58(2)	52(2)	25(1)	18(1)
3.50	*	*	*	*	194(7)	113(4)	82(3)	73(3)	52(2)	25(1)	18(1)
3.75	*	*	*	*	*	131(5)	98(4)	73(3)	52(2)	25(1)	18(1)
4.00	*	*	*	*	*	149(6)	98(4)	73(3)	65(3)	25(1)	18(1)
5.00	*	*	*	*	*	*	160(8)	115(6)	78(4)	34(2)	18(1)
6.00	*	*	*	*	*	*	*	182(11)	116(7)	43(3)	25(2)
7.00	*	*	*	*	*	*	*	*	199(14)	52(4)	25(2)

*Sample size is too large to be cost-effective for most audit applications.

NOTE: This table assumes a large population.

Table 3

Statistical Sample Results Evaluation Table for Attribute Sampling Upper Limits at Five-Percent Risk of Assessing Control Risk Too Low

Actual Number of Deviations Found

Sample Size	0	1	2	3	4	5	6	7	8	9	10
25	11.3	17.6	*	*	*	*	*	*	*	*	*
30	9.5	14.9	19.6	*	*	*	*	*	*	*	*
35	8.3	12.9	17.0	*	*	*	*	*	*	*	*
40	7.3	11.4	15.0	18.3	*	*	*	*	*	*	*
45	6.5	10.2	13.4	16.4	19.2	*	*	*	*	*	*
50	5.9	9.2	12.1	14.8	17.4	19.9	*	*	*	*	*
55	5.4	8.4	11.1	13.5	15.9	18.2	*	*	*	*	*
60	4.9	7.7	10.2	12.5	14.7	16.8	18.8	*	*	*	*
65	4.6	7.1	9.4	11.5	13.6	15.5	17.4	19.3	*	*	*
70	4.2	6.6	8.8	10.8	12.6	14.5	16.3	18.0	19.7	*	*
75	4.0	6.2	8.2	10.1	11.8	13.6	15.2	16.9	18.5	20.0	*
80	3.7	5.8	7.7	9.5	11.1	12.7	14.3	15.9	17.4	18.9	*
90	3.3	5.2	6.9	8.4	9.9	11.4	12.8	14.2	15.5	16.8	18.2
100	3.0	4.7	6.2	7.6	9.0	10.3	11.5	12.8	14.0	15.2	16.4
125	2.4	3.8	5.0	6.1	7.2	8.3	9.3	10.3	11.3	12.3	13.2
150	2.0	3.2	4.2	5.1	6.0	6.9	7.8	8.6	9.5	10.3	11.1
200	1.5	2.4	3.2	3.9	4.6	5.2	5.9	6.5	7.2	7.8	8.4

* over 20 percent

NOTE: This table presents upper limits as percentages. This table assumes a large population.

Table 4

Statistical Sample Results Evaluation Table for Attribute Sampling Upper Limits at Ten-Percent Risk of Assessing Control Risk Too Low

	Actual Number of Deviations Found										
Sample Size	**0**	**1**	**2**	**3**	**4**	**5**	**6**	**7**	**8**	**9**	**10**
20	10.9	18.1	*	*	*	*	*	*	*	*	*
25	8.8	14.7	19.9	*	*	*	*	*	*	*	*
30	7.4	12.4	16.8	*	*	*	*	*	*	*	*
35	6.4	10.7	14.5	18.1	*	*	*	*	*	*	*
40	5.6	9.4	12.8	16.0	19.0	*	*	*	*	*	*
45	5.0	8.4	11.4	14.3	17.0	19.7	*	*	*	*	*
50	4.6	7.6	10.3	12.9	15.4	17.8	*	*	*	*	*
55	4.1	6.9	9.4	11.8	14.1	16.3	18.4	*	*	*	*
60	3.8	6.4	8.7	10.8	12.9	15.0	16.9	18.9	*	*	*
70	3.3	5.5	7.5	9.3	11.1	12.9	14.6	16.3	17.9	19.6	*
80	2.9	4.8	6.6	8.2	9.8	11.3	12.8	14.3	15.8	17.2	18.6
90	2.6	4.3	5.9	7.3	8.7	10.1	11.5	12.8	14.1	15.4	16.6
100	2.3	3.9	5.3	6.6	7.9	9.1	10.3	11.5	12.7	13.9	15.0
120	2.0	3.3	4.4	5.5	6.6	7.6	8.7	9.7	10.7	11.6	12.6
160	1.5	2.5	3.3	4.2	5.0	5.8	6.5	7.3	8.0	8.8	9.5
200	1.2	2.0	2.7	3.4	4.0	4.6	5.3	5.9	6.5	7.1	7.6

* over 20 percent

NOTE: This table presents upper limits as percentages. This table assumes a large population.

APPENDIX C
SEQUENTIAL ATTRIBUTE SAMPLE SIZES AND SAMPLE RESULTS EVALUATION TABLES

Table 1

Sequential Attribute Sampling Sample Size Tables

a. Alpha Risk = .05

Beta Risk	Tolerable Deviation Rate						
	.04	.05	.06	.07	.08	.09	.10
.01	128	99	81	68	59	51	46
.02	109	84	69	58	50	44	39
.03	98	76	61	52	45	39	35
.04	90	69	56	47	41	36	32
.05	83	65	53	44	38	33	30
.06	78	61	49	41	36	31	28
.07	74	57	47	39	34	30	26
.08	70	54	44	37	32	28	25
.09	67	52	42	35	31	27	24
.10	64	50	40	34	29	26	23
Incremental Sample Size:	59	51	45	40	36	33	31

b. Alpha Risk = .10

Beta Risk	Tolerable Deviation Rate					
	.05	.06	.07	.08	.09	.10
.01			73	63	55	48
.02		75	62	53	46	41
.03		67	60	48	41	37
.04	78	62	51	44	38	34
.05	72	58	48	41	35	31
.06	68	54	45	38	33	29
.07	64	51	42	36	31	28
.08	61	49	40	34	30	26
.09	58	46	38	33	28	25
.10	56	44	37	31	27	24
Incremental Sample Size:	40	36	32	29	27	25

Table 2

Sequential Attribute Sampling
Sample Results Evaluation Table

Block	Sample Size		Number of Deviations*		
	Increment	Cumulative	Stop	Continue	Abandon
1	n_1	n_1	0	1 - 3	4
2	♦n	n_1 + ♦n	1	2 - 3	4
3	♦n	n_1 + 2♦n	2	3	4
4	♦n	n_1 + 3♦n	3	NA	4

* Cumulative
♦ Denotes incremental sample size
NA Not Applicable

APPENDIX D
TABLES FOR COMPUTING SAMPLE SIZE AND EVALUATING RESULTS FOR PPS SAMPLING

Table 1

PPS Sampling: Reliability Factors for Errors of Overstatement

Number of Overstatement Errors	Risk of Incorrect Acceptance								
	1%	5%	10%	15%	20%	25%	30%	37%	50%
0 *	4.61	3.00	2.31	1.90	1.61	1.39	1.21	1.00	0.70
1	6.64	4.75	3.89	3.38	3.00	2.70	2.44	2.14	1.68
2	8.41	6.30	5.33	4.72	4.28	3.93	3.62	3.25	2.68
3	10.05	7.76	6.69	6.02	5.52	5.11	4.77	4.34	3.68
4	11.61	9.16	8.00	7.27	6.73	6.28	5.90	5.43	4.68
5	13.11	10.52	9.28	8.50	7.91	7.43	7.01	6.49	5.68
6	14.57	11.85	10.54	9.71	9.08	8.56	8.12	7.56	6.67
7	16.00	13.15	11.78	10.90	10.24	9.69	9.21	8.63	7.67
8	17.41	14.44	13.00	12.08	11.38	10.81	10.31	9.68	8.67
9	18.79	15.71	14.21	13.25	12.52	11.92	11.39	10.74	9.67
10	20.15	16.97	15.41	14.42	13.66	13.02	12.47	11.79	10.67
11	21.49	18.21	16.60	15.57	14.78	14.13	13.55	12.84	11.67
12	22.83	19.45	17.79	16.72	15.90	15.22	14.63	13.89	12.67
13	24.14	20.67	18.96	17.86	17.02	16.32	15.70	14.93	13.67
14	25.45	21.89	20.13	19.00	18.13	17.40	16.77	15.97	14.67
15	26.75	23.10	21.30	20.13	19.24	18.49	17.84	17.02	15.67
16	28.03	24.31	22.46	21.26	20.34	19.58	18.90	18.06	16.67
17	29.31	25.50	23.61	22.39	21.44	20.66	19.97	19.10	17.67
18	30.59	26.70	24.76	23.51	22.54	21.74	21.03	20.14	18.67
19	31.85	27.88	25.91	24.63	23.64	22.81	22.09	21.18	19.67

* Use for sample size computation.

Table 2

PPS Sampling: Expansion Factors for Expected Errors

	Risk of Incorrect Acceptance								
	1%	5%	10%	15%	20%	25%	30%	37%	50%
Factor	1.9	1.6	1.5	1.4	1.3	1.25	1.2	1.15	1.0

APPENDIX E
VARIABLES SAMPLING TABLES

Table 1: Table of Z Values and Corresponding Confidence Levels

Table 2: Ratio of Desired Allowance for Sampling Risk to Tolerable Error

Table 1

Table of Z Values and Corresponding Confidence Levels

Z Value	Confidence Level
1.00	68.26
1.038	70.00
1.282	80.00
1.645	90.00
1.96	95.00
2.00	95.44
2.33	98.00
2.5	98.76
2.576	99.00
2.8	99.48

Table 2

Ratio of Desired Allowance for Sampling Risk to Tolerable Error

Risk of Incorrect Acceptance	Risk of Incorrect Rejection			
	.20	.10	.05	.01
.01	.355	.413	.457	.525
.025	.395	.456	.500	.568
.05	.437	.500	.543	.609
.075	.471	.532	.576	.641
.10	.500	.561	.605	.668
.15	.511	.612	.653	.712
.20	.603	.661	.700	.753
.25	.653	.708	.742	.791
.30	.707	.756	.787	.829
.35	.766	.808	.834	.868
.40	.831	.863	.883	.908
.45	.907	.926	.937	.952
.50	1.000	1.000	1.000	1.000

GLOSSARY

Acceptance Sampling —Term used to describe the *one sided upper precision limit* approach in attribute sampling.

Allowance for Sampling Risk — Same as *precision.*

Alpha Risk—A component of sampling risk. The risk of assessing control risk too high leading to inefficiency (tests of controls) or the risk of incorrect rejection (substantive testing).

Analytical Procedures — Substantive tests that involve the study and comparison of plausible relationships among data.

Attribute — Evidence of compliance with a prescribed control procedure.

Attribute Sampling — Sampling in tests of controls.

Audit Risk —The risk that the auditor may fail to modify an opinion on financial statements that are materially misleading.

Basic Precision — An assumed amount of possible unknown or undiscovered error in the population when sampling.

Beta Risk — A component of sampling risk. The risk of assessing control risk too low leading to ineffectiveness (tests of controls) or the risk of incorrect acceptance (substantive testing).

Block Selection —A sample selection method in which the auditor applies audit procedures to all items or transactions that occurred in the same block (a group of consecutive transactions) of time or sequence of transactions.

Confidence Interval —The range of likely values within which the true population value should lie. Also called precision interval.

Confidence Level —The degree of belief the auditor has in the obtained results.

Control Procedure — Policies and procedures prescribed by management to provide assurance that specific objectives will be met.

Control Risk — The risk that the internal control structure will fail to prevent or detect a material error from occurring in the financial statements.

Detection Risk — The risk that the auditor's substantive tests will fail to detect a material error in the financial statements.

Deviation —The absence of evidence that a prescribed control procedure was applied.

Difference Estimation — A variables sampling technique used to estimate the total error in the population when the auditor has both the recorded book value and audited value for each item in the sample.

Discovery Sampling — A statistical sampling approach used to "discover" at least one deviation if the percentage of deviations in the population is at or above a specified level. Commonly used in fraud auditing.

Expected Error — An estimate of the dollar error that exists in a population.

Fixed Attribute Sampling Plan —A statistical sampling plan used to test the effectiveness of a prescribed control procedure.

Haphazard Selection — A nonstatistical sample selection method in which the auditor selects the sample items without intentional bias to include or exclude items in the population.

Inherent Risk — The risk of material error in a transaction class or account balance assuming there are no internal controls.

Internal Control Structure —The policies and procedures established by management to provide reasonable assurance that objectives will be met.

Judgment Selection — A nonstatistical sample selection method in which the sample is selected based on the auditor's sound and seasoned judgment.

Logical Unit —A balance or transaction that encompasses a sampling unit in a PPS sample.

Lower Precision Limit —In attribute sampling, the minimum error estimated to exist in the population at a specified confidence level. In variables sampling, the minimum estimated value of the population at a specified confidence level. Used to define the lowest value in a precision interval.

Mean —The expected value of an item in a population or sample.

Mean Per Unit Estimation — A variables sampling technique used to estimate the total dollar amount of the population by calculating a sample mean of the audited values and multiplying it by the number of items in the population.

Measures of Location — A representative value often used to describe a population or sample data. The three measures of location include the mean, median, and mode.

Median —The middle value in a population or sample that has been ordered from smallest to largest.

Mode —The most frequently occurring value in a population or sample.

Nonsampling Risk —The risk that the auditor will fail to detect a material control weakness or misstatement due to factors other than sampling.

Population — The total collection of items about which a conclusion will be expressed.

Precision — Measures how close the sample estimate is to the true population value.

Precision Interval —The range of likely values within which the true population value should lie. Same as confidence interval.

Probability-Proportional-to-Size (PPS) Sampling — A variation of attribute sampling that is useful in detecting the dollar amount of overstatement when only a few errors are expected. Used for substantive audit testing.

Range — The difference between the lowest and highest values in the population or sample.

Ratio Estimation — A variables sampling technique used to estimate the total dollar amount of a population by calculating the ratio between the audited and book values in the sample and using this ratio to make the estimate.

Reliability — A term used in auditing to mean confidence level.

Risk of Assessing Control Risk Too High —The risk that the auditor assesses control risk too high, thereby causing the sample results to reject the planned reliance on internal control when in fact the true population compliance rate supports such reliance. This relates to the efficiency of the audit. Also called *risk of underreliance.*

Risk of Assessing Control Risk Too Low — The risk that the auditor assesses control risk too low, thereby causing the sample results to support the planned reliance on internal control when in fact the true population deviation rate does not support such reliance. This relates to the effectiveness of the audit. Also called *risk of overreliance.*

Sample —The collection of sampling units drawn from the sample frame that will be subjected to auditing procedures.

Sample Frame — A listing of the total sampling units.

Sampling Interval — The population divided by the sample size. This number is always rounded down.

Sampling Risk —The risk that the conclusions reached by the auditor based on an analysis of a sample may differ from those conclusions that would be reached by examining the entire population.

Sampling Unit —The individual items that make up a population.

Sequential Attribute Sampling (stop-or-go-sampling) — A sampling plan in which the sample is conducted in several steps. After each step, the auditor must decide whether to stop, continue, or abandon the sample.

Simple Random Sampling — A method of selecting a sample such that each item in the population has an equal chance of being selected.

Standard Deviation — A measurement of the variation about the mean value in a population or sample.

Standard Error of the Mean —The standard deviation of a sampling distribution.

Stratified Sampling — A sampling method in which the population of interest is divided into several sub-populations called strata. A sample is selected from each stratum.

Substantive Testing — Tests used to evaluate the fair presentation of dollar amounts that appear in the financial statements.

Systematic Sampling — A method of selecting samples such that sampling units are selected at fixed intervals in a population.

Tainting —The proportion of error in a logical unit in a probability-proportional-to-size (PPS) sample. Expressed as a percentage.

Tests of Controls — Procedures designed to provide knowledge about the effectiveness of internal control procedures and policies in preventing or detecting material misstatements.

Tolerable Deviation Rate —The maximum number of deviations from the prescribed control procedure that can be tolerated without altering planned audit testing.

Tolerable Error — The auditor's assessment of materiality with respect to the population being audited.

Upper Error Limit — The maximum error estimated to exist in the population at a specified confidence level. Used in PPS sampling.

Upper Precision Limit — In attribute sampling, the maximum error estimated to exist in the population at a specified confidence level. In variables sampling, the maximum estimated value of the population at a specified confidence level. Used to define the highest value in a precision interval.

Variables Sampling — Used when the auditor wishes to estimate a total dollar amount of transactions or balances in substantive testing.

Variance —The value of the standard deviation squared.

Z Value — Tells the auditor how many standard deviations the value of interest is from the mean.

SUGGESTED REFERENCES

AICPA. 1981. *Statement on Auditing Standards No. 39, Audit Sampling.* New York: American Institute of Certified Public Accountants.

AICPA. 1983. *Audit Sampling.* Audit and Accounting Guide. New York: American Institute of Certified Public Accountants.

AICPA. 1983. *Statement on Auditing Standards No. 47, Audit Risk and Materiality in Conducting an Audit.* New York: American Institute of Certified Public Accountants.

AICPA. 1988. *Statement on Auditing Standards No. 55, Consideration of Internal Control Structure in a Financial Statement Audit.* New York: American Institute of Certified Public Accountants.

Apostolou, Barbara. 1991. *Sampling for Internal Auditors: A Self-Study Course.* Altamonte Springs, Fla: The Institute of Internal Auditors, Inc. *

Arens, Alvin A. and James K. Loebbecke. 1981. *Applications of Statistical Sampling to Auditing.* Prentice-Hall, Inc.

Arkin, Herbert. 1982. *Sampling Methods for the Auditor: An Advanced Treatment.* Prentice-Hall, Inc. *

Arkin, Herbert. 1984. *Handbook of Sampling for Auditing and Accounting, Third Edition.* Prentice-Hall, Inc. *

Bailey, Andrew D., Jr. 1981. *Statistical Auditing.* New York: Harcourt Brace Jovanovich, Inc.

Brightman, Harvey J. 1986. *Statistics in Plain English.* Cincinnati: South-Western Publishing Co.

Kazmier, Leonard J. 1988. *Business Statistics 2/ed.* Schaum's Outline Series in Business. McGraw-Hill Book Company.

Roberts, Donald M. 1978. *Statistical Auditing.* New York: American Institute of Certified Public Accountants.

Stringer, Kenneth W. and Trevor R. Stewart. 1986. *Statistical Techniques for Analytical Review and Auditing.* John Wiley & Sons, Inc. *

* Available directly from The Institute of Internal Auditors.

SUGGESTED REFERENCES

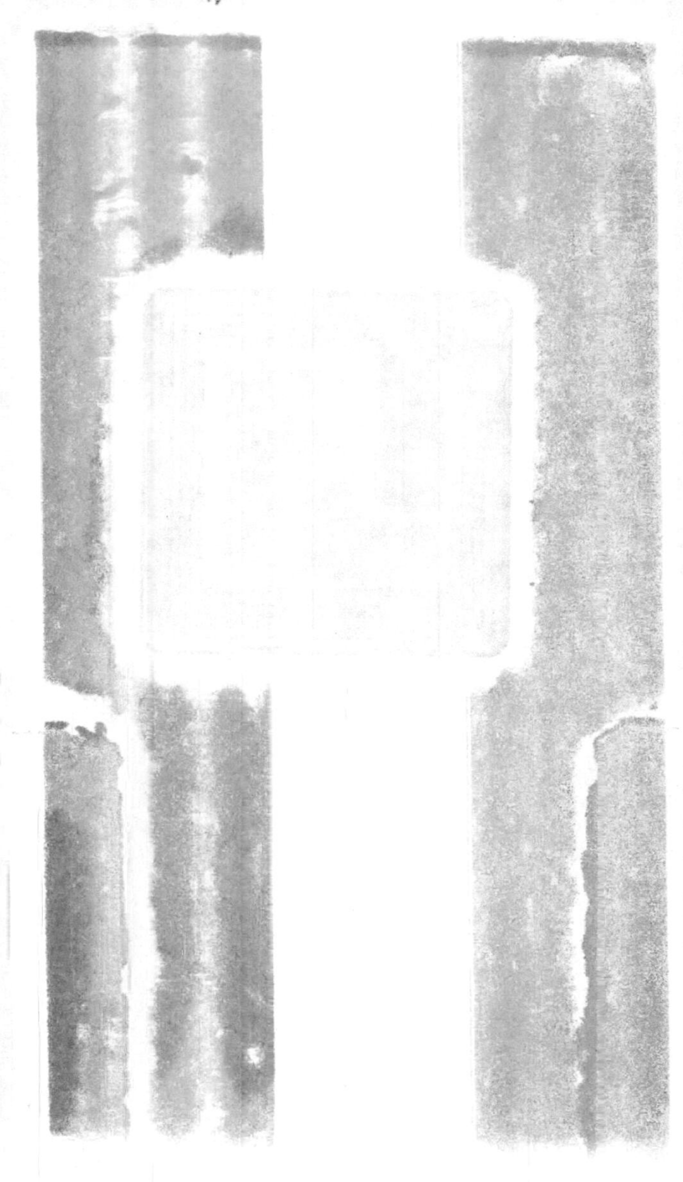

3 1542 00163 8232

